DYNAMICAL SEARCH

APPLICATIONS OF DYNAMICAL SYSTEMS IN SEARCH AND OPTIMIZATION

DYNAMICAL SEARCH

APPLICATIONS OF DYNAMICAL SYSTEMS IN SEARCH AND OPTIMIZATION

Interdisciplinary Statistics

Luc Pronzato
CNRS Université de Nice Sophia-Antipolis, France

Henry P. Wynn
University of Warwick,
Coventry, United Kingdom

Anatoly A. Zhigljavsky
University of Cardiff, United Kingdom

CHAPMAN & HALL/CRC

Boca Raton London New York Washington, D.C.

Library of Congress Cataloging-in-Publication Data

Pronzato, Luc, 1959–.
Dynamical search : applications of dynamical systems in search and optimisation / Luc Pronzato, Henry P. Wynn, Anatoly A. Zhigljavsky.
p. cm.
Includes bibliographical references and index.
ISBN 0-8493-0336-2
1. Search theroy. 2. Differentiable dynamical systems. I. Wynn, Henry P. II. Zhigliavakiĭ, A. A.(Anatoliĭ Aleksandrovich)
III. Title.
T57.97.P76 1999
003—dc21 99-32954
CIP

International Standard Book Number 0-8493-0336-2
Library of Congress Card Number 99-32954
Printed in the United States of America 1 2 3 4 5 6 7 8 9 0
Printed on acid-free paper

To our parents,
to Jan,
and our friends.

Preface

If a two-author, two-nationality book is a complex operation, the reader may perhaps imagine the difficulties of completing this three-author work and the level of gratitude which is due to families, friends, funding bodies and our publisher for their support and patience. It had to compete with all the usual events of modern life such as emigration, marriage and job change, to name the more pleasant, and was also completed during a period of upheaval in Eastern Europe which affected personal contact at certain times. Five centres were involved: the CNRS laboratory in Sophia-Antipolis, for Luc, City University and then the University of Warwick, for Henry and the University of St. Petersburg and then the University of Cardiff, for Anatoly. To all our colleagues we are very grateful for their help and encouragement.

A remarkable level of support and hospitality was given to us by friends and families. Carole Lassalle and her now not-so-young daughters Salomé and Morgane, Isabelle Penod, Harald Keller and others made life in Antibes an ideal mixture of work and play. Henry's wife Jan, sons Hamish and Robin and the extended Wynn-Baldwin clan, including Fionna Dunlop, broke away from otherwise lively times in London to give us patient encouragement. Anatoly's family and St. Petersburg friends were very supportive on visits to the city and the hospitality continues in Cardiff.

Academic friends, some of whom are also co-workers on other projects, should get a special mention for the sort of osmotic support which they would probably be too modest to admit to, particulary Eric Walter, Giovanni Pistone, Dan Naiman, Peter Caines, Pavel Kargaev, and also Rainer Schwabe, Anthony Atkinson, Werner Muller, Peter Hackl, Valery Fedorov and all the other MODA Conference colleagues.

Certain people gave specific help and advice: Max Chekmasov did some early programming work; Brigitte Vallée, Caroline Series,

Martin Huxley and Glyn Harman gave advice on Farey sequences, continued fractions and aspects of dynamical systems.

We are very grateful for support from several grants. A UK Engineering and Physical Sciences (EPSRC) Visiting Fellowship supported Anatoly for an early visit to City University and the work has also been covered by other EPSRC grants under both mathematics and engineering headings. Luc and Henry shared a British Council Alliance grant and Anatoly was partly supported by CNRS itself and partly by a "PAST" grant from the French Ministry of Education during important visits to France.

Finally, many thanks to Chapman and Hall-CRC for staying with us till the end on what at times may have seemed a somewhat speculative venture and one which certainly saw many deadlines come and go.

L.P., H.P.W., A.A.Z.

Contents

Notation

Typography

X, x, α	scalars
$\boldsymbol{x}, \boldsymbol{\alpha}$	column vectors
$\boldsymbol{X}$	matrix
$\boldsymbol{x}^T, \boldsymbol{X}^T$	transposed of $\boldsymbol{x}$ and $\boldsymbol{X}$
$\mathcal{X}$	set
$[\boldsymbol{x}]_i$	i-th component of $\boldsymbol{x}$
$[\boldsymbol{X}]_{ij}$	(i, j)-th entry of $\boldsymbol{X}$

Common symbols and acronyms

$\lfloor x \rfloor$	integer part of x
$\{x\}$	fractional part of x
$\mathbf{1}$	vector with all entries equal to 1
$\mathbf{0}$	null vector
$\boldsymbol{O}$	null matrix
$[A_k, B_k]$	search interval (non-normalised) at iteration k of a line-search algorithm
$\mathcal{B}(\boldsymbol{x}, \rho)$	ball with centre $\boldsymbol{x}$ and radius ρ
$\mathcal{C}_k$	consistency set at iteration k
e	exponential, $\log \mathsf{e} = 1$, $\mathsf{e} \simeq 2.71828$
e_k, e'_k	normalised values of E_k and E'_k, $e_k = (E_k - A_k)/(B_k - A_k)$, $e'_k = (E'_k - A_k)/(B_k - A_k)$
$\mathsf{E}_x\{\cdot\}$	expected value with respect to x
$\mathsf{E} \log \mathrm{L}_n$	$\mathsf{E}_{\boldsymbol{x}_1}\{\log L_n(\boldsymbol{x}_1)\}$ when $\boldsymbol{x}_1$ has some distribution in $\mathcal{R}_1$

$\mathbf{E}\mathrm{L}_n^{\gamma}$	$\mathbf{E}_{\boldsymbol{x}_1}\{L_n^{\gamma}(\boldsymbol{x}_1)\}$ when $\boldsymbol{x}_1$ has some distribution in $\mathcal{R}_1$
(E_k, E_k')	k-th pair of testing points for a second-order line-search algorithm
$\mathcal{E}(\boldsymbol{c}, \boldsymbol{M})$	ellipsoid with centre $\boldsymbol{c}$ and expansion matrix $\boldsymbol{M}$, $\mathcal{E}(\boldsymbol{c}, \boldsymbol{M}) = \{\boldsymbol{x} \in \mathbb{R}^n \ / \ (\boldsymbol{x} - \boldsymbol{c})^T \boldsymbol{M}^{-1}(\boldsymbol{x} - \boldsymbol{c}) \leq 1\}$
$\mathcal{E}_o^*(\mathcal{C})$	minimum-volume outer ellipsoid for $\mathcal{C}$
$\bar{\mathcal{E}}_o^*(\mathcal{C})$	minimum-volume outer ellipsoid for $\mathcal{C}$ with fixed centre
$\mathcal{E}_i^*(\mathcal{C})$	maximum-volume inner ellipsoid for $\mathcal{C}$
$\bar{\mathcal{E}}_i^*(\mathcal{C})$	maximum-volume inner ellipsoid for $\mathcal{C}$ with fixed centre
$f(\cdot)$	objective function
H_γ	Rényi entropy of order γ
$\boldsymbol{I}$ ($\boldsymbol{I}_n$)	identity matrix ($n \times n$ identity matrix)
$\mathtt{I}_{\mathcal{A}}(\cdot)$	indicator function of the set $\mathcal{A}$
$\mathtt{int}(\mathcal{X})$	interior of the set $\mathcal{X}$
L_k	length of the search interval at iteration k, $L_k = B_k - A_k$, or volume of search region $\mathcal{R}_k$
$\bar{\mathrm{L}}_n^{1-\alpha}$	$\sup\{t \ / \ \mathtt{Pr}(L_n(\boldsymbol{x}_1) \geq t) > \alpha\}$
$\mathtt{M}\mathrm{L}_n$	$\sup_{\boldsymbol{x}_1 \in \mathcal{R}_1} L_n(\boldsymbol{x}_1)$
$\mathtt{N}_n$	number of different regions $\mathcal{R}_n(\boldsymbol{x}^*)$ for $\boldsymbol{x}^* \in \mathcal{R}_0$
$\mathcal{P}, \mathcal{P}_N$	partitions
$\mathtt{Pr}(\cdot)$	probability
$P_n(\beta)$	$\mathtt{Pr}[(1/n) \log L_n(\boldsymbol{x}_1) < \beta]$
r_k	convergence rate at iteration k
R	ergodic convergence rate
$\mathcal{R}$	base region for renormalisation
$\mathcal{R}_k, k \geq 0$	search region (not normalised)
$T(\cdot)$	transformation defining a dynamical system
$\mathtt{vol}(\mathcal{R})$	volume of the region $\mathcal{R}$
W_1	$-\lim_{n\to\infty}(1/n)\mathbf{E}\log \mathrm{L}_n$
W_2	$-\lim_{n\to\infty}(1/n)\log \mathbf{E}\mathrm{L}_n$
W_γ	$\frac{1}{1-\gamma}\lim_{n\to\infty}(1/n)\log \mathbf{E}\mathrm{L}_n^{\gamma-1}$, $\gamma \neq 1$, $\gamma \geq 0$
W_∞	$-\lim_{n\to\infty}(1/n)\log \mathtt{M}\mathrm{L}_n$
x^* or $\boldsymbol{x}^*$	search object, in $\mathbb{R}$ or $\mathbb{R}^d$
δ^n	relative accuracy in the minimisation of $f(\cdot)$
ϑ_d	volume of the d-dimensional unit ball
$\lambda_{\min}(\boldsymbol{X})$	minimal eigenvalue of the matrix $\boldsymbol{X}$
$\lambda_{\max}(\boldsymbol{X})$	maximal eigenvalue of the matrix $\boldsymbol{X}$
Λ	Lyapunov exponent

μ	measure
μ_T	invariant measure for the dynamical system defined by $T(\cdot)$
$\mu_{\mathcal{L}}$	Lebesgue measure
$\boldsymbol{\Pi}$	transition matrix for a Markov chain
$\bar{\boldsymbol{\pi}}$	vector of invariant probabilities for a Markov chain
$\bar{\boldsymbol{\pi}}^{(1)}$	vector of initial probabilities for a Markov chain
ϱ	log-rate ($\varrho = -\log R$)
φ	Golden-Section ratio, $\varphi = (\sqrt{5}-1)/2 \simeq 0.61803398875$
ϕ	invariant density

Convention for derivatives

$\frac{\partial \boldsymbol{X}(\alpha)}{\partial \alpha}$	derivative of a matrix $\boldsymbol{X}$ with respect to the scalar α, $\left[\frac{\partial \boldsymbol{X}(\alpha)}{\partial \alpha}\right]_{ij} = \frac{\partial [\boldsymbol{X}]_{ij}(\alpha)}{\partial \alpha}$
$\nabla f(\boldsymbol{x})$	gradient vector of $f(\cdot)$ with respect to $\boldsymbol{x}$, $[\nabla f(\boldsymbol{x})]_i = \frac{\partial f(\boldsymbol{x})}{\partial [\boldsymbol{x}]_i}$
$\nabla^2 f(\boldsymbol{x})$	Hessian matrix of $f(\cdot)$ with respect to $\boldsymbol{x}$, $\left[\nabla^2 f(\boldsymbol{x})\right]_{ij} = \frac{\partial^2 f(\boldsymbol{x})}{\partial [\boldsymbol{x}]_i \partial [\boldsymbol{x}]_j}$
$\frac{\partial f(\boldsymbol{X})}{\partial \boldsymbol{X}}$	derivative matrix of $f(\cdot)$ with respect to $\boldsymbol{X}$, $\left[\frac{\partial f(\boldsymbol{X})}{\partial \boldsymbol{X}}\right]_{ij} = \frac{\partial f(\boldsymbol{X})}{\partial [\boldsymbol{X}]_{ij}}$
$\boldsymbol{J}_T(\boldsymbol{x})$	Jacobian matrix of the transformation $T(\cdot) : \boldsymbol{x} \to T(\boldsymbol{x})$, $[\boldsymbol{J}_T(\boldsymbol{x})]_{ij} = \frac{\partial [T(\boldsymbol{x})]_i}{\partial [\boldsymbol{x}]_j}$

CHAPTER 1

Introduction

This book covers a branch of what would historically have been called Operational Research. Today it should more properly be called an interface subject between two or perhaps more disciplines. The title was thus deliberately chosen to reflect the positioning between dynamical systems and search theory. The first of these is now a very broad area covering parts of nonlinear mathematics, ergodic theory and the still-fashionable chaos theory. Search theory, on the other hand, is harder to pin down as a coherent discipline, and the bulk of the subject is probably in computer science and artificial intelligence, where it does real work in producing solutions to logical questions in the retrieval of information, and is studied mathematically under the heading of algorithmic complexity. Search also has roots in optimisation theory: for example, line search algorithms are built into many optimisation algorithms. It is this second aspect of search which dominates this work, and hence the inclusion of optimisation in the subtitle. The connection with recent theories in computer science and the study of algorithms arises because the link with dynamical systems, and ergodic theory in particular, introduces stochasticity into algorithms in a new way.

The germ of the idea for the research programme leading to this book was the following: a number of classical algorithms in optimisation, which appear to converge smoothly to the target, behave in a haphazard fashion when looked at a local, or second order, level. This is clear from the following, and perhaps simplest, example. The binary expansion of an irrational number between 0 and 1 converges to that number. However, each successive digit may occur in a seemingly haphazard fashion (the link with normal numbers in number theory is well known). Now this binary expansion can be linked in an obvious way to a bifurcation search algorithm in which successive intervals surrounding the number are halved. The intervals converge in a smooth way, and yet the behaviour of the digits appears locally random. The well-known Golden Section algorithm, which is the limiting case of the Fibonacci search algorithm, for a local extremum of a function, can be shown to behave in a similar way after a little more analysis. Both binary expansions and the Golden-Section itself are well known to be associated with particular dynamical systems.

The link between certain types of search algorithms and dynamical systems can fortunately be made in a formal way using the following idea. Imagine some algorithm in which there is a generic type of set containing the target, with a sequence of this type of set converging to the target. Line search is a simple example. Another important example is the ellipsoidal algorithm in linear programming, where ever-decreasing ellipses are generated which contain and converge to the target. One can also imagine multi-dimensional algorithms where the basic set is a hyper-rectangle, a kind of generalisation of line search. The idea is that, after each iteration, one transforms the current region back to some standard version, for example, in line search to a standard interval, for the ellipsoidal algorithm to a standard sphere and in the imagined algorithm to a unit hyper-cube. In performing this transformation the target itself will have to move, because the shifting and rescaling represented in renormalisation will also be applied to the target. In a broad class of cases, this moving target can be seen to be the trajectory of a point under a dynamical system.

The dynamical system associated with the algorithm can be used to study the behaviour of the algorithm, and even to improve it. Working in a dynamical system environment also has suggested new types of algorithms, for example the "hyper-cube algorithm" hinted at above. In particular, the rates of convergence of the original algorithm can be translated into various exponents and entropies of the associated dynamical systems. We feel that perhaps the most important aspect of the book is precisely the fact that the types of entropies arising in this way are not the most studied in dynamical systems. The contraction rate of the set in the algorithm corresponds to the rate of expansion of the dynamical system. For example, the Rényi entropy with the exponent $\gamma = 2$, rather than the usual Shannon entropy ($\gamma = 1$), arises naturally as a measure of contraction of a set.

Making a link between dynamical systems and search theory may be considered worthwhile in itself, but the goal of improving existing algorithms, and suggesting new ones, was a big motivation for us. One way of improving an algorithm is to carry out some kind of relaxation in which the algorithm is pushed towards higher rates of convergence. For example, for the ellipsoid algorithm, the so-called deep cut method leads to some bizarre behaviour by the associated dynamical system. Even without relaxation, the classical steepest descent algorithm reveals a complex fractal structure after renormalisation and, with relaxation, bifurcation and chaotic properties appear.

One result of the approach is that some algorithms classically considered as optimal are in fact optimal in the worst case, and, ergodically, the worst case events have measure zero. It is thus possible to construct algorithms with improved ergodic performances. Using a Bayesian approach, with a prior distribution for the search object, we also show that algorithms with improved asymptotic expected performances can be constructed. The improvements rely on a particular property of the problem: search for the root of a linear function, for the minimum of a unimodal symmetric function, for the solution of a linear-programming problem with d constraints in dimension d, and so on. Indeed, it is the particular assumption about the problem that gives an association of a time-invariant dynamical system with the search algorithm. It often happens that the assumption is only *locally* satisfied around the solution. However, since the algorithm is convergent, it will ultimately reach a region where the assumption is more and more reasonable. For that reason, the asymptotic behaviour of the algorithm coincides with that of the time-invariant dynamical system, even in situations where the assumption is only valid locally.

An important step consists in deriving algorithms with good performances for a finite (and small) number of iterations from algorithms with good ergodic performances. This derivation is detailed for second-order line-search algorithms in Chapter 5. A key tool for that is to place the algorithm in initial conditions such that the support of the prior distribution for the search object coincides with the support of the invariant density for the associated dynamical system. Indeed, it often happens that the location of the search object that corresponds to the worst cases for the performances of the algorithm lies outside the support of this invariant density.

A pleasing connection is made between the Gauss map of continued fractions, and the associated Farey map and Farey tree, and a special sub-class of line-search algorithms of which the Golden-Section and Fibonacci search are special cases. These are so-called symmetric algorithms. As in many fields one is never too far from number theory. We hope again that the search motivation and use of Rényi entropies to define rates of approximation gives some freshness to the approach.

Consistency is considered in Chapter 2, and renormalisation of consistency regions in Chapter 3. The first technical issue is to link the rate of convergence of the algorithm with the characteristics of the dynamical process. This is considered in Chapter 4. Line-search algorithms are considered in Chapter 5, ellipsoidal algorithms in Chapter 6. A second issue is to study more closely the behaviour of algorithms that only exhibit convergence to a point attractor when renormalisation is not used. For instance, Chapter 7 concerns the case of the steepest-descent algorithm. The approach allows a careful analysis of the behaviour of this classical algorithm en route to the attractor. A link between algorithms and dynamical systems is then exploited to improve rates of convergence.

CHAPTER 2

Consistency

In many deterministic problems in which the objective is to discover or approximate some unknown quantity, one is presented with incomplete, or partial, information. The technical problem is then to make maximum use of this information, in order to learn as much as possible about the unknown quantity. In such situations, a simple rule is invaluable: the truth is whatever is consistent with the data.

Information in search problems is presented in a variety of forms, but most typically in the value of a function at a selected point, or site. From this we learn about the function, or some aspects of it, such as its minimum. Put another way, the observation forces a restriction on the function. All subsequent information, and, in the sequential case, all subsequent choice of observation site, takes this restriction into account. As we take more and more observations, the restriction becomes stronger and stronger, and confines the target quantity in a tighter and tighter region, which we typically call the *consistency region*. One of the most fascinating aspects of the idea of consistency is that the geometry of the consistency region is often of considerable mathematical interest. Even when, as we shall see for line search, it is a simple interval, the manner of its construction, or updating from observation to observation, will be of great interest. In addition, although there will be a smallest such region —the one which makes maximum use of current information— this may be difficult to construct. In that case, we may take an easier region containing the "true" consistency region, and hence, of course, the target.

The usefulness of the consistency idea cannot be understated. In conducting the research which led to this work, on many occasions issues were clarified by simply asking carefully what is the true consistency region? Should we take a larger region? What will be the effect on convergence? Will the dynamic behaviour be easier to study as a result? References to "optimistic" consistency regions (using more information) and "pessimistic" (using less) were especially useful.

A final introductory remark is that many, if not most, algorithms in areas such as artificial intelligence, computational learning and computational geometry use the notion of consistency, sometimes without a clear reference to the fact. This may be because it is so basic an idea that it is taken for granted. It is certainly underplayed in statistics, where stochastic methods such as Bayesian methods tend to allow a wide consistency region with the search effort being carried by posterior distributions.

2.1 Consistency in discrete search

Many search problems can be represented as quadruples $\{\Theta, \mathcal{X}, f, \mathcal{Y}\}$ where $\Theta = \{\theta\}$ is a parameter set, that is, a collection of all possible parameter values θ, $\mathcal{X} = \{x\}$ is a test set, that is, a collection of all possible test elements, and $f : \mathcal{X} \times \Theta \mapsto \mathcal{Y}$ is a *search function* mapping $\mathcal{X} \times \Theta$ to some space $\mathcal{Y}$, which is assumed finite or countable: $\mathcal{Y} = \{y_1, \ldots, y_m\}$, $m \leq \infty$. A value $f(x, \theta)$ for fixed $x \in \mathcal{X}$ and $\theta \in \Theta$ is regarded as a test, or experimental result at a test element x, when the unknown parameter is θ; see O'Geran, Wynn and Zhigljavsky (1993).

A *non-sequential design* $\mathcal{D}_n$ of length n is a collection of n test elements $x_1, \ldots, x_n$, also called design points, chosen before observation starts. When the true parameter is θ, testing at design points $x_1, \ldots, x_n$ yields the data $[f(x_1, \theta), \ldots, f(x_n, \theta)] \in \mathcal{Y}^n$. A partition of Θ can thus be associated with any design $\mathcal{D}_n = \{x_1, \ldots, x_n\}$:

$$\mathcal{P}_N : \Theta = \bigcup_{(y_1,\ldots,y_n)} \Theta_{(y_1,\ldots,y_n)}, \tag{2.1}$$

where $y_k \in \mathcal{Y}$ for every $k = 1, \ldots, n$ and

$$\Theta_{(y_1,\ldots,y_n)} = \{\theta \in \Theta \; / \; f(x_1, \theta) = y_1, \ldots, f(x_n, \theta) = y_n\}$$

is the consistency set, that is, the set of parameters *consistent* with the data $y_1, \ldots, y_n$. In many important cases the information used is not in the raw y_i but in some filtered version, such as comparisons between y_i.

The *consistency sets* Θ_i, where $i = (y_1, \ldots, y_n) \in \mathcal{Y}^n$, can also be considered as uncertainty regions for θ so that all $\theta \in \Theta_i$ lead to the same observations. With this in mind, we shall call two parameter values θ, θ' non-distinguishable if they fall in the same consistency set: $\theta, \theta' \in \Theta_i$ for some i. The performances of a design are usually measured in terms of precision on the location of θ. We show below that uncertainty on θ can be measured through the entropy of the partition.

Entropy

Assume that $(\Theta, \mathcal{B}, \mu)$ is a probability space with probability measure μ, which defines the volume of subsets in Θ. In later sections we shall consider μ as a formal prior distribution on Θ.

Let

$$\mathcal{P}_N : \Theta = \bigcup_{i=1}^{N} \Theta_i\,, \quad \Theta_i \cap \Theta_j = \emptyset \quad \text{for } i \neq j,\ \ N \leq \infty\,, \tag{2.2}$$

be a partition of Θ. For a given θ, the volume $L(\theta) = p_i = \mu(\Theta_i)$ of the region Θ_i that contains θ is a natural characteristic of uncertainty. The average volume of the uncertainty region is therefore

$$\mathbf{E}_\theta\{L(\theta)\} = \sum_i p_i^2\,,$$

which is related to the second-order Rényi entropy of the partition:

$$H_2(\mathcal{P}_N, \mu) = -\log \sum_{i=1}^{N} p_i^2\,.$$

More generally, for any real γ

$$\mathbf{E}_\theta\{L^{\gamma-1}(\theta)\} = \sum_i p_i^\gamma\,,$$

and therefore for any $\gamma \neq 1$

$$H_\gamma(\mathcal{P}_N, \mu) = \frac{1}{1-\gamma} \log \mathbf{E}_\theta\{L^{\gamma-1}(\theta)\}\,,$$

where

$$H_\gamma(\mathcal{P}_N, \mu) = \frac{1}{1-\gamma} \log \sum_{i=1}^{N} p_i^\gamma$$

is the Rényi entropy of order γ; see Section 8.1.

Shannon entropy corresponds to the average logarithm of the volume of the uncertainty region:

$$H_1(\mathcal{P}_N, \mu) = -\sum_{i=1}^{N} p_i \log p_i = \mathbf{E}_\theta\{-\log L(\theta)\}\,.$$

The topological entropy $H_0(\mathcal{P}_N, \mu)$ simply counts the number of subsets in the partition, that is $H_0(\mathcal{P}_N, \mu) = \log N$.

As mentioned in Chapter 1, from a practical point of view, the average volume of the uncertainty region is often more justified as a quality criterion for a given design than the average of the logarithm of the volume of the uncertainty region. Let us give further arguments in favour of the second-order Rényi entropy of the partition $\mathcal{P}_N$ generated by a given design.

(i) *Repeatability.* Assume that θ, θ' are independently distributed with the prior measure μ. Then the probability of θ, θ' falling in the same subset of the partition (2.1) can be written as

$$\Pr\{\theta, \theta' \text{ are non-distinguishable}\} = \sum_i p_i^2 .$$

(ii) *Pair-splitting.* Let Θ be a finite set with $|\Theta| = n$, and $\mathcal{D}_N$ be a design that produces the partition

$$\Theta = \bigcup_{i=1}^{N} \Theta_i, \quad |\Theta_i| = n_i, \quad n_1 + \cdots + n_N = n, \quad p_i = n_i/n .$$

Then the number of non-distinguishable pairs (θ, θ') after application of the design $\mathcal{D}_N$ equals

$$\sum_{i=1}^{N} \frac{n_i(n_i+1)}{2} = \frac{n^2}{2} \sum_i p_i^2 + \frac{n}{2},$$

and is thus related to $\sum p_i^2$. The numerical solution of various discrete search problems such as tree-search, screening and group-testing, Mastermind, etc., see O'Geran, Wynn and Zhigljavsky (1991, 1993), has shown that using the pair-splitting as the optimality criterion for the selection of new test points often leads to more economical sequential designs than using Shannon entropy.

2.2 Line-search algorithms

2.2.1 Root-finding and bifurcation

Let $f(\cdot)$ be a continuous function on $[A_0, B_0)$ such that $f(A_0) \leq 0 < f(B_0)$ and x^* be the unique root of $f(\cdot)$ in (A_0, B_0). We consider *first-order* algorithms based on evaluations of the sign of $f(\cdot)$ and comparisons of function values. To emphasize, the values of $f(\cdot)$ are not used but only certain filtered information expressed as

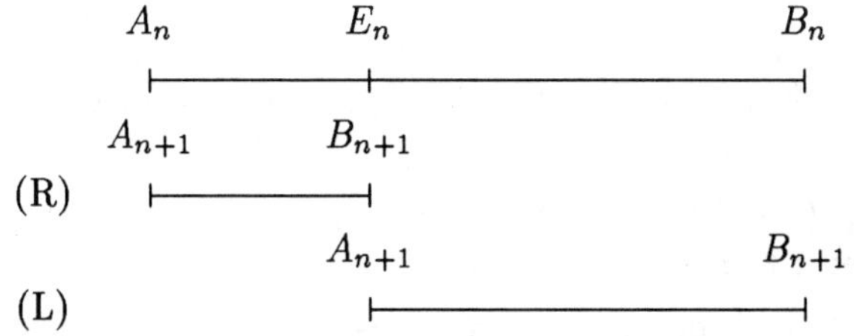

Figure 2.1 *One iteration in first-order line search*

comparisons or similar entities. Let $\mathcal{C}_n = [A_n, B_n)$ denote the uncertainty (or consistency) interval at iteration n and $E_n \in (A_n, B_n)$ be a test-point where the sign of $f(\cdot)$ is observed. If $f(E_n) \leq 0$, then necessarily $x^* \in [E_n, B_n)$ and $[A_{n+1}, B_{n+1}) = [E_n, B_n)$ can be considered as the next consistency interval $\mathcal{C}_{n+1}$; see Figure 2.1. Similarly, if $f(E_n) > 0$, then we can take $[A_{n+1}, B_{n+1}) = [A_n, E_n)$. In Figure 2.1, (R) and (L) respectively stand for deletion of the right and left side of the interval.

Independently of the choice of E_n in (A_n, B_n) we then have for all n: $f(A_n) \leq 0 < f(B_n)$. Note that were $f(E_n) = 0$ to be observed, then x^* is found and the algorithm can be stopped. Assuming that sign(0) is positive, this situation can be neglected.

A first-order line-search algorithm is thus defined by two choices:

(i) initial uncertainty interval $[A_1, B_1) \supseteq [A_0, B_0)$,

(ii) selection rule for the test-point $E_n \in [A_n, B_n)$, $n \geq 0$.

The *bifurcation algorithm* corresponds to $[A_1, B_1) = [A_0, B_0)$ and $E_n = (A_n + B_n)/2$. It will be considered again in Section 3.2.1.

A family of algorithms, considered in more detail in Sections 3.2.1 and 5.1, is obtained by choosing E_n as follows:

$$E_n = \begin{cases} A_n + e(B_n - A_n) & \text{if } |f(A_n)| \leq |f(B_n)| \\ A_n + (1-e)(B_n - A_n) & \text{otherwise} \end{cases} \tag{2.3}$$

with $e \leq 1/2$. The bifurcation algorithm corresponds to the special case $e = 1/2$. An algorithm in this family can be formally described in the following form (we assume that $f(A_0)$ and $f(B_0)$ have opposite signs).

Algorithm 1. (General first-order line-search algorithm.)

Step (0) Choose $N \geq 1$ and/or $\delta > 0$ to define the stopping rule of Step (ii). Take $[A_1, B_1) = [A_0, B_0)$. If $f(A_0) > 0 > f(B_0)$ then change $f(\cdot)$ into $-f(\cdot)$.

Step (i) At iteration n, choose E_n according to (2.3). If $f(E_n) \leq 0$, take $[A_{n+1}, B_{n+1}) = [E_n, B_n)$, otherwise take $[A_{n+1}, B_{n+1}) = [A_n, E_n)$.

Step (ii) If $n+1 \geq N$ and/or $B_{n+1} - A_{n+1} \leq \delta$ stop, otherwise set n to $n+1$ and go to Step (i).

We do not consider the case where $[A_1, B_1)$ is taken larger than $[A_0, B_0)$. The reason for sometimes expanding the initial search interval will be explained later in the case of second-order line-search algorithms.

The reduction, or convergence, rate at iteration n is defined as $r_n = L_{n+1}/L_n$, where $L_k = B_k - A_k$ is the length of the uncertainty interval at iteration k, so that

$$L_n = L_0 \prod_{i=0}^{n-1} r_i \, ;$$

see Chapter 4.

2.2.2 Continued fraction expansion

Any irrational number x^* in $[0, 1)$ has a unique continued fraction expansion of the form

$$x^* = [a_1, a_2, \ldots] = \cfrac{1}{a_1 + \cfrac{1}{a_2 + \cfrac{1}{a_3 + \cdots}}} \, ,$$

where the *partial quotients* $a_1, a_2, \ldots$ are positive integers. If we terminate the expansion at iteration n, we obtain the *convergents* of the continued fraction, which give approximations of x^*. More generally, define

$$Q_n(\epsilon) = [a_1, a_2, \ldots, a_n, \epsilon] = \cfrac{1}{a_1 + \cfrac{1}{a_2 + \cdots \cfrac{1}{a_n + \epsilon}}} \, .$$

Then there is a unique ϵ_{n+1} such that

$$Q_n(\epsilon_{n+1}) = x^* \, .$$

The continued fraction expansion for a rational number x^* is finite and nonunique:

$$x^* = [a_1, a_2, \ldots, a_n] = \begin{cases} [a_1, a_2, \ldots, a_n - 1, 1] & \text{if } a_n > 1 \, , \\ [a_1, a_2, \ldots, a_{n-1} + 1] & \text{if } a_n = 1 \, . \end{cases}$$

Note that one can also write (see Schoisengeier, 1990)

$$x^* = [a_1, a_2, \ldots, a_n] = [a_1, a_2, \ldots, a_n, \infty]\,. \tag{2.4}$$

If x^* is an irrational number in $(0,1)$,

$$\begin{array}{ll} Q_n(0) < x^* < Q_n(1) & (n \text{ even}) \\ Q_n(1) < x^* < Q_n(0) & (n \text{ odd}) \end{array}$$

Since ϵ can achieve any value in $[0,1)$, these are exactly the consistency intervals for x^* based on partial information $a_1, a_2, \ldots, a_n$, and can be summarized as $\{Q_n(\epsilon)\,,\ 0 < \epsilon < 1\}$.

We can develop many of the special properties of these intervals from the classical theory of continued fractions, for which one can refer to Rockett and Szüsz (1992). First define the following system:

$$\begin{aligned} \epsilon_1(x^*) &= x^* \\ \epsilon_2(x^*) &= \left\{\frac{1}{\epsilon_1(x^*)}\right\} = \frac{1}{\epsilon_1(x^*)} - \left\lfloor\frac{1}{\epsilon_1(x^*)}\right\rfloor \\ &\vdots \\ \epsilon_{n+1}(x^*) &= \left\{\frac{1}{\epsilon_n(x^*)}\right\} = \frac{1}{\epsilon_n(x^*)} - \left\lfloor\frac{1}{\epsilon_n(x^*)}\right\rfloor \end{aligned} \tag{2.5}$$

where $\{\cdot\}$ is the fractional part and $\lfloor\cdot\rfloor$ is the integer part. For convenience, we shall simply denote $\epsilon_n(x^*) = \epsilon_n$.

Returning to the convergents, we can express $Q_n(\epsilon)$ as a ratio of terms unambiguously by simply clearing the fractions in the continued fraction:

$$Q_n(\epsilon) = \frac{p_n + \epsilon p_n'}{q_n + \epsilon q_n'}\,,$$

and for x^*:

$$x^* = Q_n(\epsilon_{n+1}) = \frac{p_n + \epsilon_{n+1} p_n'}{q_n + \epsilon_{n+1} q_n'}\,, \tag{2.6}$$

for some integers p_n, q_n, p_n', q_n'. Using the identities

$$\begin{aligned} Q_n(\epsilon) &= Q_{n-1}\left(\frac{1}{a_n + \epsilon}\right), \\ Q_{n-1}(\epsilon) &= Q_{n-2}\left(\frac{1}{a_{n-1} + \epsilon}\right), \end{aligned}$$

clearing the fractions and equating coefficients in numerators and denominators, we obtain a number of important identities:

$$\begin{aligned} p_n &= a_n p_{n-1} + p_{n-2}\,, \\ q_n &= a_n q_{n-1} + q_{n-2}\,, \end{aligned} \tag{2.7}$$

with $p_{-1} = 1$, $q_{-1} = 0$, $p_0 = 0$, $q_0 = 1$. We also have $p'_n = p_{n-1}$ and $q'_n = q_{n-1}$. The end points of the consistency region are thus given by $(n \geq 0)$

$$\begin{aligned} Q_n(0) &= \frac{p_n}{q_n}\,, \\ Q_n(1) &= \frac{p_n + p_{n-1}}{q_n + q_{n-1}}\,. \end{aligned}$$

One can check by induction that

$$p_n q_{n-1} - p_{n-1} q_n = (-1)^{n-1}\,, \quad n \geq 0\,. \tag{2.8}$$

Moreover, a_n can be explicitly expressed in terms of ϵ_n:

$$a_n = \left\lfloor \frac{1}{\epsilon_n} \right\rfloor\,.$$

We are now in a position to fully describe the updating of the consistency region. Consider the case when n is odd, which implies

$$Q_{n-1}(0) = \frac{p_{n-1}}{q_{n-1}} < Q_n(0) = \frac{p_n}{q_n}\,.$$

Then, in the interval $[Q_{n-1}(0), Q_n(0)]$, we have

$$Q_{n-1}(0) < Q_n(1) < Q_{n+1}(0) < Q_{n+1}(1) < Q_n(0)\,,$$

or, in terms of explicit values:

$$\frac{p_{n-1}}{q_{n-1}} < \frac{p_n + p_{n-1}}{q_n + q_{n-1}} < \frac{p_{n+1}}{q_{n+1}} < \frac{p_{n+1} + p_n}{q_{n+1} + q_n} < \frac{p_n}{q_n}\,.$$

We shall return to a full analysis of the process as an approximation algorithm in Sections 3.2.2 and 5.2. Just to fix terminology, we shall refer to

$$\left[\frac{p_{n-1}}{q_{n-1}}, \frac{p_n}{q_n}\right]\,, \quad \left[\frac{p_{n+1}}{q_{n+1}}, \frac{p_n}{q_n}\right]\,,$$

as *pessimistic* intervals and the true consistency sets

$$\left[\frac{p_n + p_{n-1}}{q_n + q_{n-1}}, \frac{p_n}{q_n}\right]\,, \quad \left[\frac{p_{n+1}}{q_{n+1}}, \frac{p_{n+1} + p_n}{q_{n+1} + q_n}\right]\,,$$

as the *optimistic* intervals.

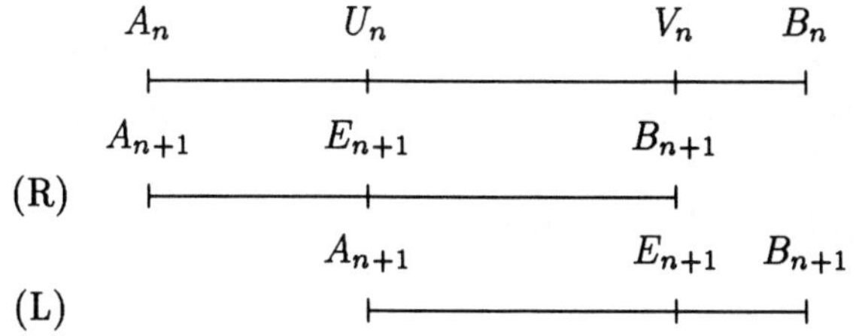

Figure 2.2 *One iteration in second-order line search*

2.2.3 Optimisation based on comparisons of function values

We consider the minimisation of a uniextremal function $f(\cdot)$ on a given interval $[A_0, B_0)$ using a *second-order* algorithm, as defined by Kiefer (1957). Let x^* be the unknown point at which $f(\cdot)$ is minimum: $f(\cdot)$ is decreasing for $x \leq x^*$ and non-decreasing for $x > x^*$ (or non-increasing for $x \leq x^*$ and increasing for $x > x^*$).

At iteration n we compare the values of $f(\cdot)$ at two points U_n and V_n in the current uncertainty interval $C_n = [A_n, B_n)$, with $U_n < V_n$. Then, after this iteration if $f(U_n) \geq f(V_n)$ we delete $[A_n, U_n)$; otherwise we delete $[V_n, B_n)$. Note that, in a practical implementation of the algorithm, both $[A_n, U_n)$ and $[V_n, B_n)$ can be deleted in the case where $f(U_n) = f(V_n)$, but the algorithm should then be re-initialised. (This will not be considered here because it has no effect on the performance characteristics that are considered.) The remaining part of the interval defines the uncertainty interval $[A_{n+1}, B_{n+1})$ for the next iteration; see Figure 2.2. In this figure, (R) and (L) again stand respectively for Right and Left deletion. In each case, one of the two points U_n, V_n is carried forward to $[A_{n+1}, B_{n+1})$. Let E_{n+1} denote this point. At iteration $n+1$ we compare $f(E_{n+1})$ to the value of $f(\cdot)$ at a new point E'_{n+1}.

A second-order line-search algorithm is therefore defined by the choice of the

(i) initial uncertainty interval $[A_1, B_1) \supseteq [A_0, B_0)$,

(ii) initial test-point $E_1 \in [A_1, B_1)$,

(iii) selection rule for E'_{n+1}, $n \geq 0$.

It can formally be described as follows.

Algorithm 2. (General second-order line-search algorithm.)

Step (0) Choose $N \geq 2$ and/or $\delta > 0$ to define the stopping rule of Step (iii).

Step (i) Define an interval $[A_1, B_1)$ such that $[A_0, B_0) \subseteq [A_1, B_1)$, select a point $E_1 \in (A_1, B_1)$, and set $n = 1$.

Step (ii) At iteration n select a point $E'_n \in (A_n, B_n)$, define $U_n = \min\{E_n, E'_n\}$, $V_n = \max\{E_n, E'_n\}$, compare the values of $f(\cdot)$ at U_n and V_n. Then, if $f(U_n) \geq f(V_n)$ delete the segment $[A_n, U_n)$; otherwise delete $[V_n, B_n)$. The remaining part of the interval defines the uncertainty interval $[A_{n+1}, B_{n+1})$ for the next iteration. Either U_n or V_n belongs to $[A_{n+1}, B_{n+1})$; denote this point E_{n+1}.

Step (iii) If $n + 1 \geq N$ and/or $B_{n+1} - A_{n+1} \leq \delta$ stop; otherwise set n to $n + 1$ and go to Step (ii).

One important aspect of this general type of algorithm is that the choice of $[A_1, B_1) \supseteq [A_0, B_0)$, which corresponds to an expansion of the initial uncertainty interval, affects the initialisation of the dynamical system to be presented in Section 5.7.1 and has a strong influence on some performance characteristics; see Section 5.7.2.

When $\epsilon > 0$, that is when $[A_1, B_1)$ is strictly larger than $[A_0, B_0)$, it may happen, for instance for the window algorithm presented below, that $E'_n < A$ or $E'_n > B$, whereas $f(\cdot)$ is only defined on $[A, B]$. Then we make a linear extension of $f(\cdot)$: define $f(E'_n) = f(A) + A - E'_n$ for $E'_n < A$ and $f(E'_n) = f(B) + E'_n - B$ for $E'_n > B$.

Analogously to the case of the first-order line-search algorithms, the reduction rate at iteration n is defined as $r_n = L_{n+1}/L_n$, where $L_k = B_k - A_k$ is the length of the uncertainty interval at iteration k, so that

$$L_n = L_0 \prod_{i=0}^{n-1} r_i \, .$$

For *symmetric algorithms*, E'_n in Step (ii) of Algorithm 2 is selected according to the rule $E'_n = A_n + B_n - E_n$. In that case, the length L_n does not depend on the sequence of (R) and (L) deletions and is thus independent of the objective function $f(\cdot)$. It only depends on E_1. Figure 2.3 presents L_n/L_1 as a function of E_1 for $n = 2, \ldots, 7$.

The most famous second-order line-search algorithms are the Fibonacci and the Golden-Section (GS) algorithms. Both are symmetric.

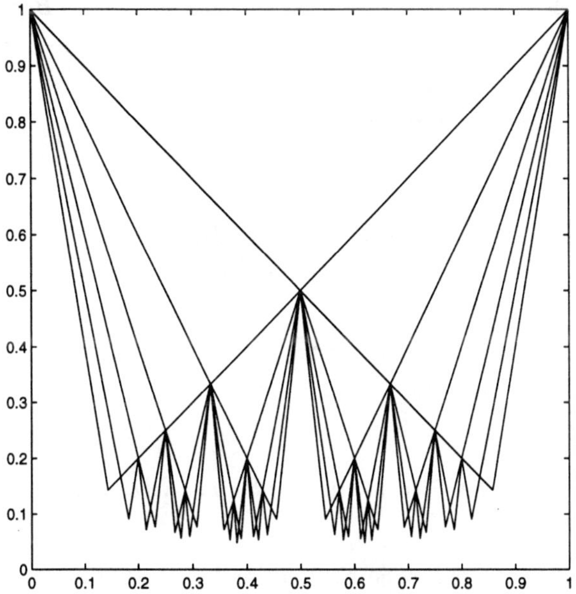

Figure 2.3 $L_n(E_1)/L_1$ *as a function of* E_1, $n = 2, \ldots, 7$, *for a symmetric algorithm*

Golden-Section and Fibonacci algorithms

The Golden-Section (GS) algorithm is defined by

$$[A_1, B_1) = [A_0, B_0)\,, \quad E_1 = A_1 + \varphi L_1\,,$$

$$E'_n = \begin{cases} A_n + \varphi L_n & \text{if} \quad E_n = A_n + (1-\varphi)L_n \\ A_n + (1-\varphi)L_n & \text{if} \quad E_n = A_n + \varphi L_n \end{cases} \tag{2.9}$$

where $L_n = B_n - A_n$ and where $\varphi = (\sqrt{5}-1)/2 \simeq 0.61804$ is the largest root of $\varphi^2 + \varphi - 1 = 0$ and is called the Golden-Section ratio. The key property of the algorithm is that E_{n+1} satisfies

$$\frac{E_{n+1} - A_{n+1}}{L_{n+1}} \in \{1-\varphi, \varphi\} \quad \forall n \geq 0\,. \tag{2.10}$$

This algorithm is known to be asymptotically worst-case optimal in the class of all uniextremal functions; see Kiefer (1957) and Du and Hwang (1993), Theorem 9.2.2, p. 181. The convergence rate at iteration n satisfies $r_0 = 1$ and $r_n = \varphi$, $\forall n \geq 1$, so that $L_n = L_0\varphi^{n-1}$. When the number of function evaluations is fixed,

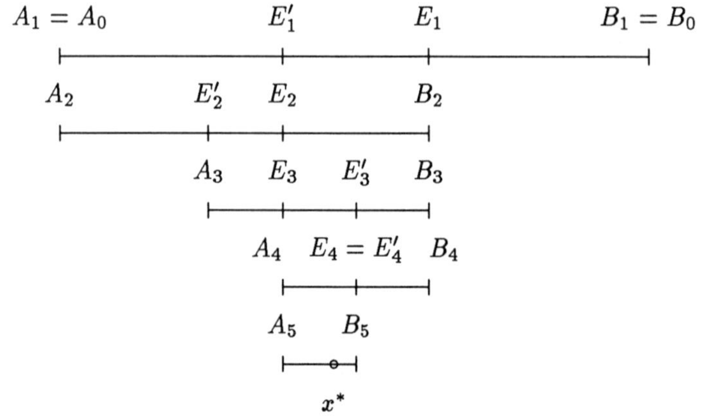

Figure 2.4 *Five iterations of the Fibonacci algorithm*

say equal to N, the worst-case optimal algorithm in the sense of L_N is the Fibonacci algorithm (Kiefer, 1957), for which

$$[A_1, B_1) = [A_0, B_0)\,, \quad E_1 = A_1 + \frac{F_N}{F_{N+1}} L_1\,,$$

$$E'_n = A_n + B_n - E_n\,,$$

where $(F_i)_{i=1}^{\infty} = \{1, 1, 2, 3, 5, 8, 13, \ldots\}$ is the Fibonacci sequence, defined by $F_1 = F_2 = 1$ and $F_n = F_{n-1} + F_{n-2}, n > 2$. The algorithm satisfies

$$r_0 = 1\,, \quad r_n = \frac{F_{N+1-n}}{F_{N+2-n}}\,, \quad L_n = L_1 \frac{F_{N+2-n}}{F_{N+1}}\,, \quad 1 \leq n \leq N\,;$$

see Figure 2.4. Note that in the last iteration one assumes that the two test-points coincide. In practice they should be chosen as close as possible. The algorithm stops at $n = N$, so that one can define $r_n = 1$ and $L_n = L_N$ for $n \geq N$.

From the recurrence defining F_n, we obtain

$$F_n = \frac{(1+\varphi)^n - (-\varphi)^n}{\sqrt{5}}\,. \tag{2.11}$$

When N tends to infinity, the ratio of L_N for the GS algorithm to L_N for the Fibonacci algorithm tends to $(1+\varphi)^2/\sqrt{5} \simeq 1.17082$.

Algorithms based on section-invariant numbers

An ordered collection of m numbers $\mathcal{U} = \{v_1, \ldots, v_m\}$ is called *section-invariant* if $0 < v_1 < \cdots < v_m < 1$ and there exists an associated collection of m numbers $\mathcal{U}' = \{v'_1, \ldots, v'_m\}$ in $(0,1)$ such that for every $j = 1, \ldots, m$

$$R(v_j, v'_j) \in \mathcal{U} \text{ and } L(v_j, v'_j) \in \mathcal{U}\,,$$

where

$$\begin{aligned} L(v,v') &= \tfrac{\max(v,v')-\min(v,v')}{1-\min(v,v')}\,, \\ R(v,v') &= \tfrac{\min(v,v')}{\max(v,v')}\,; \end{aligned} \tag{2.12}$$

see Section 8.3 for details.

We use these section-invariant numbers as the basis for the construction of a family of second-order line-search algorithms. In a suitable sense, some of these algorithms have better convergence rates than the Fibonacci and Golden-Section algorithms, which are optimal in the worst-case sense (for finite N and as $N \to \infty$, respectively).

Let us specialize Algorithm 1 to the case when the selection rule for E'_n is based on the use of section-invariant numbers.

Algorithm 3.

Let $f(\cdot)$ be the function to be minimised on $[A_0, B_0)$ and $\mathcal{U} = \{v_1, \ldots, v_m\}$, $\mathcal{U}' = \{v'_1, \ldots, v'_m\}$ be the sets defining the section-invariant numbers v_i.

Step (0) Choose $j \in \{1, \ldots, m\}$ and $\epsilon \geq 0$ used at Step (i) as well as N and/or δ to be used for the stopping rule of Step (iii).

Step (i) Compute $A_1 = A_0 - \epsilon L_0$, $B_1 = B_0 + \epsilon L_0$. Take $E_1 = A_1 + v_j(B_1 - A_1)$ and set $n = 1$.

Step (ii) Compute $E'_n = A_n + v'_i L_n$ if $E_n = A_n + v_i L_n$. Define $U_n = \min\{E_n, E'_n\}$, $V_n = \max\{E_n, E'_n\}$, compare the values of $f(\cdot)$ at U_n and V_n. Then, if $f(U_n) \geq f(V_n)$ delete the segment $[A_n, U_n)$; otherwise delete $[V_n, B_n)$. Either U_n or V_n belongs to $[A_{n+1}, B_{n+1})$; denote this point E_{n+1}.

Step (iii) If $n + 1 \geq N$ and/or $B_{n+1} - A_{n+1} \leq \delta$ stop; otherwise set n to $n + 1$ and go to Step (ii).

A key point in the algorithm is the generalisation of the property (2.10) of the GS algorithm:

$$\frac{E_{n+1} - A_{n+1}}{L_{n+1}} \in \mathcal{U} \quad \forall n \geq 0\,.$$

The GS algorithm corresponds to $\epsilon = 0$ and the pair $(\mathcal{U},\mathcal{U}')$ given by $\mathcal{U} = \{\varphi, 1-\varphi\}$, $\mathcal{U}' = \{1-\varphi, \varphi\}$. This algorithm is the fastest, according to criteria of Chapter 4, within the class above restricted to $|\mathcal{U}| = 2$. When $|\mathcal{U}| \leq 3$, the algorithm with the best ergodic rate of convergence is defined by Table 8.4 of Section 8.3 (see Examples 5.8 and 8.1). However, it is only marginally superior to the GS algorithm. When $|\mathcal{U}| \leq 4$, the best algorithm in the class above with respect to many ergodic and finite-sample criteria is defined by $\epsilon = (1-a)/2 \simeq 0.40294$, $E_1 = A_1 + bL_1$ and

$$\mathcal{U} = \{a, b, 1-b, 1-a\}\,,\ \mathcal{U}' = \{a', 1-b, b, 1-a'\}\,,$$

where

$$b = 2a^3 - 4a^2 + 3a\,,\ a' = 2a - a^2\,,\quad \text{and } a \simeq 0.19412$$

is the smallest positive root of the polynomial $2t^4 - 8t^3 + 11t^2 - 7t + 1$. We call this algorithm GS4; see Section 5.7.1 for a detailed description. We shall also consider the algorithm obtained by taking $\epsilon = 0$ in GS4; we shall call it $\mathrm{GS4}_0$ algorithm.

Midpoint algorithm

The algorithm is defined by $[A_1, B_1) = [A_0, B_0)$, $E_1 = A_1 + e_1 L_1$ with e_1 any irrational number in $[0, 1)$, and

$$E'_n = \frac{A_n + B_n}{2}\,,\ \forall n \geq 1\,.$$

Window algorithm

The algorithm is defined by

$$\begin{aligned} [A_1, B_1) &= [A_0 - \epsilon L_0, B_0 + \epsilon L_0)\,, \\ E_1 &= (A_1 + B_1)/2 - wL_1/2\,, \end{aligned} \tag{2.13}$$

and

$$E'_n = \begin{cases} E_n + wL_n & \text{if } E_n < \frac{1}{2}(A_n + B_n) \\ E_n - wL_n & \text{otherwise,} \end{cases} \tag{2.14}$$

where $\epsilon \geq 0$ and $w > 0$ are tuning parameters. The window width, defined by the ratio $|E'_n - E_n|/L_n$, is thus fixed and equals w. The values of ϵ and w could be chosen optimally for each N and each performance characteristic of Chapter 4. A rather exhaustive analysis for reasonable values of N $(10 \leq N \leq 30)$ has led to the choice $\epsilon = 0.3772$ and $w = 0.15$, which is close to optimality for all criteria.

Note that $r_0 > 1$ for $\epsilon > 0$, which is the case for the GS4 and window algorithms. The reason for expanding the interval $[A_0, B_0)$ will be explained later, see Section 5.7.2.

2.3 Consistency in linear and convex programming

Linear and convex programming provide a rich environment for the dynamical process approach of this book. The following sections lay the foundation for a fuller development in Chapter 6. It is essential to understand how the consistency sets change geometrically and their contraction rates for different kinds of algorithms.

2.3.1 Linear programming

We shall consider the following linear programming problem:

$$\text{minimise} \quad \boldsymbol{c}^T \boldsymbol{x} \tag{2.15}$$

$$\text{subject to} \quad \boldsymbol{A}\boldsymbol{x} \leq \boldsymbol{b}, \tag{2.16}$$

where $\leq$ means componentwise inequality.

Note that the standard form of a linear programming problem is

$$\text{minimise} \quad \boldsymbol{c}^T \boldsymbol{x}$$

$$\text{subject to} \quad \boldsymbol{C}\boldsymbol{x} = \boldsymbol{d}, \quad \boldsymbol{x} \geq 0;$$

see, *e.g.*, Minoux (1983). Equality constraints are not considered here because we require feasible sets with nonempty interiors. When equality constraints are present, they should thus be eliminated, for example by eliminating some variables and thus reducing the dimension of the problem, or by using an exact penalisation of the objective. Assume that $\boldsymbol{x}$ should satisfy (2.16) together with $\boldsymbol{C}\boldsymbol{x} = \boldsymbol{d}$. Then, the equality constraints are replaced by the additional inequality constraints $\boldsymbol{C}\boldsymbol{x} - \boldsymbol{d} \leq \boldsymbol{\alpha}$, $-\boldsymbol{C}\boldsymbol{x} + \boldsymbol{d} \leq \boldsymbol{\alpha}$, and the objective (2.15) is modified into $\boldsymbol{c}^T \boldsymbol{x} + \beta \sum_i \alpha_i$, with β a large positive number and α_i the i-th component of $\boldsymbol{\alpha}$.

The evaluation of $\boldsymbol{c}^T \boldsymbol{x}$ at the feasible point $\boldsymbol{x}^{(k)}$ at iteration k introduces the additional information that the optimum point $\boldsymbol{x}^*$ satisfies $\boldsymbol{c}^T \boldsymbol{x}^* \leq \boldsymbol{c}^T \boldsymbol{x}^{(k)}$. Let $\mathcal{C}_k$ be the consistency set obtained after the evaluation of $\boldsymbol{c}^T \boldsymbol{x}^{(1)}, \ldots, \boldsymbol{c}^T \boldsymbol{x}^{(k)}$,

$$\mathcal{C}_k = \{\boldsymbol{x} \in \mathbb{R}^d \ / \ \boldsymbol{A}\boldsymbol{x} \leq \boldsymbol{b}, \ \boldsymbol{c}^T \boldsymbol{x} \leq \boldsymbol{c}^T \boldsymbol{x}^{(n)}, n = 1, \ldots, k\}. \tag{2.17}$$

Choosing $\boldsymbol{x}^{(k+1)}$ in the interior of $\mathcal{C}_k$ allows reduction of the size of

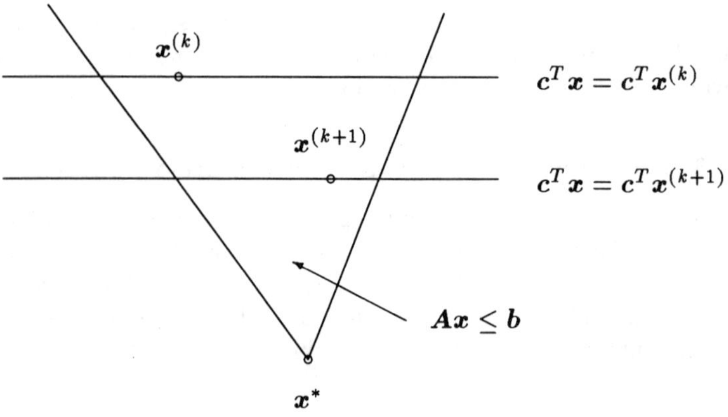

Figure 2.5 *Consistency sets in linear programming*

$\mathcal{C}_k$, that is to obtain a new consistency set $\mathcal{C}_{k+1} \subset \mathcal{C}_k$, see Figure 2.5. Note that the $\mathcal{C}_k$'s are polyhedras in $\mathbb{R}^d$. A key relation is that between the reduction of the size of the sets $\mathcal{C}_k$ and the decrease of the objective function $c^T x$. This is considered in the next section for general convex functions.

2.3.2 Convex programming using gradient information: cut-off methods

We consider the problem of minimizing $f(\cdot)$, a convex function of $x \in \mathbb{R}^d$, subject to constraints $Ax \leq b$. Let $\nabla f(x)$ be the gradient of $f(\cdot)$ at x. (When $f(\cdot)$ is not everywhere differentiable, subgradients can be used instead; see, *e.g.*, Shor (1985).) From the convexity of $f(\cdot)$, any solution x^* of the optimisation problem satisfies

$$f(x^*) \geq f(x) + (x^* - x)^T \nabla f(x) \quad \forall x \in \mathbb{R}^d .$$

Let $\mathcal{C}_k$ be the consistency set obtained after the evaluation of the first k gradients $\nabla f(x^{(1)}), \ldots, \nabla f(x^{(k)})$:

$$\begin{aligned} \mathcal{C}_k &= \{x \in \mathbb{R}^d \ / \ Ax \leq b , \\ &\quad f(x^*) \geq f(x^{(n)}) + (x^* - x^{(n)})^T \nabla f(x^{(n)}) \, , n = 1, \ldots, k\} . \end{aligned}$$

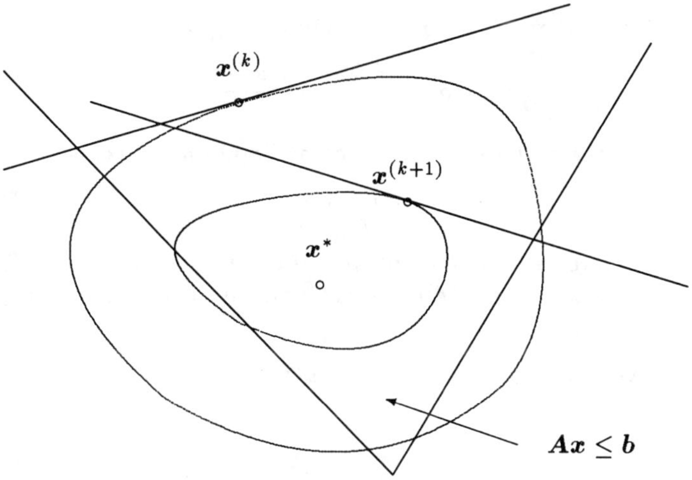

Figure 2.6 *Consistency sets in convex programming*

Choosing $x^{(k+1)}$ in the interior of $\mathcal{C}_k$, we obtain a new consistency region $\mathcal{C}_{k+1} \subset \mathcal{C}_k$.

When $f(x^*)$ is unknown, we can use the fact that $f(x^*) \leq f(x^{(n)})$ $(n = 1, \ldots, k)$ to construct the consistency sets

$$\mathcal{C}_k = \{x \in \mathbb{R}^d \ / \ Ax \leq b, (x^* - x^{(n)})^T \nabla f(x^{(n)}) \leq 0 \ , \ n = 1, \ldots, k\} \, ;$$

see Figure 2.6. Again, the sets $\mathcal{C}_k$ are polyhedra in $\mathbb{R}^d$.

The point $x^{(k+1)}$ should be chosen to ensure a fast decrease of $f(\cdot)$. In the *cutting-plane* method (Kelley, 1960), $x^{(k+1)}$ minimises a linear approximation of $f(\cdot)$ constructed from previous evaluations of $f(\cdot)$ and its gradient $\nabla f(\cdot)$. It is obtained by solving the linear programming problem:

$$\begin{aligned} &\text{minimise} && \xi \\ &\text{subject to} && Ax \leq b \\ & && (x - x^{(n)})^T \nabla f(x^{(n)}) + f(x^{(n)}) \leq \xi \ , \ n = 1 \ldots, k \, . \end{aligned}$$

The convexity of $f(\cdot)$ can be used to relate the accuracy of the minimisation in terms of $f(\cdot)$ to the precision of the location of x^* in terms of size of $\mathcal{C}_k$. Define

$$\Delta = \max_{x \in \mathcal{C}_0} f(x) - f(x^*)$$

and the relative accuracy in the minimisation of $f(\cdot)$

$$\delta^k = \frac{f(\hat{\boldsymbol{x}}^k) - f(\boldsymbol{x}^*)}{\Delta}, \tag{2.18}$$

with $\hat{\boldsymbol{x}}^k$ the best value of $\boldsymbol{x}$ found at iteration k, that is

$$\hat{\boldsymbol{x}}^k = \arg \min_{n=1,\ldots,k} f(\boldsymbol{x}^{(n)}).$$

To compare the size of $\mathcal{C}_k$ to that of $\mathcal{C}_0$, we use the value $\vartheta(\mathcal{C}_k)$ of the largest homothety coefficient for which the translated image of $\mathcal{C}_0$ can be contained in $\mathcal{C}_k$:

$$\forall \mathcal{C} \subseteq \mathcal{C}_0\,,\ \vartheta(\mathcal{C}) = \max_{\boldsymbol{u} \in \mathbb{R}^d} \{z \in \mathbb{R}^+ \ / \ \boldsymbol{u} + z\mathcal{C}_0 \subseteq \mathcal{C}\}. \tag{2.19}$$

One can show that

$$\delta^k \leq \vartheta(\mathcal{C}_k). \tag{2.20}$$

Indeed, consider the transformation

$$\boldsymbol{x} \in \mathcal{C}_0 \mapsto F(\boldsymbol{x}) = \boldsymbol{x}^* + \delta^k(\boldsymbol{x} - \boldsymbol{x}^*).$$

From the convexity of $f(\cdot)$, for any $\boldsymbol{x} \in \mathcal{C}_0$ and any $\boldsymbol{z}$ on the segment joining $\boldsymbol{x}$ to $\boldsymbol{x}^*$

$$\frac{f(\boldsymbol{z}) - f(\boldsymbol{x}^*)}{||\boldsymbol{z} - \boldsymbol{x}^*||} \leq \frac{f(\boldsymbol{x}) - f(\boldsymbol{x}^*)}{||\boldsymbol{x} - \boldsymbol{x}^*||},$$

and thus, for $\boldsymbol{z} = F(\boldsymbol{x})$:

$$\frac{f[F(\boldsymbol{x})] - f(\boldsymbol{x}^*)}{\delta^k ||\boldsymbol{x} - \boldsymbol{x}^*||} \leq \frac{f(\boldsymbol{x}) - f(\boldsymbol{x}^*)}{||\boldsymbol{x} - \boldsymbol{x}^*||}.$$

This implies

$$\begin{aligned} f[F(\boldsymbol{x})] &\leq [f(\hat{\boldsymbol{x}}^k) - f(\boldsymbol{x}^*)] \frac{f(\boldsymbol{x}) - f(\boldsymbol{x}^*)}{\max_{\boldsymbol{x} \in \mathcal{C}_0} f(\boldsymbol{x}) - f(\boldsymbol{x}^*)} + f(\boldsymbol{x}^*) \\ &\leq f(\hat{\boldsymbol{x}}^k), \end{aligned}$$

and thus $F(\boldsymbol{x}) \in \mathcal{C}_k$, which gives (2.20).

From the definition (2.19) of $\vartheta(\mathcal{C})$ we then obtain

$$\vartheta(\mathcal{C}_k) \leq \left(\frac{L_k}{L_0}\right)^{1/d},$$

where L_k denotes $\mathtt{vol}(\mathcal{C}_k)$, and thus from (2.20)

$$\delta^k \leq \left(\frac{L_k}{L_0}\right)^{1/d}.$$

This property is used to show that choosing $x^{(k+1)}$ as the center of gravity of the consistency set $\mathcal{C}_k$, which corresponds to the so-called *method of centroids*, gives an optimal method (in the worst case) for convex programming in terms of the number of iterations, in the class of methods which use only local information about the behaviour of $f(\cdot)$; see Levin (1965). We define the rate of convergence at iteration k by

$$r_k = \frac{L_{k+1}}{L_k} = \frac{\mathtt{vol}(\mathcal{C}_{k+1})}{\mathtt{vol}(\mathcal{C}_k)}.$$

For the method of centroids the rate satisfies

$$r_k \leq 1 - \left(1 - \frac{1}{d+1}\right)^d,$$

and thus $r_k \leq 1 - 1/\mathsf{e} \simeq 0.63212$, with $\log \mathsf{e} = 1$, whatever the dimension d; see Minoux (1983). This can be compared, for example, to the rate (2.24) of the outer-ellipsoid algorithm with central cuts, see Sections 2.3.4, 3.3.1, 3.3.2 and Figure 3.10. From this we can compute the number $N(\epsilon)$ of iterations required to get a precision ϵ for δ^n, which is such that $\delta^n \leq \epsilon$ for $n \geq N(\epsilon)$, namely

$$N(\epsilon) \leq -\frac{d\log(\frac{1}{\epsilon})}{\log(1-\frac{1}{\mathsf{e}})}.$$

This implies

$$N(\epsilon) \leq 2.19\, d\log(1/\epsilon). \tag{2.21}$$

However, the determination of the center of gravity of $\mathcal{C}_k$ is not easy (no algorithm with polynomial complexity is known). Another method, optimal too in terms of number of iterations, is based on the construction of inner ellipsoids with maximum volumes and is presented in Section 2.3.4.

2.3.3 Set covering for global optimisation

Consistency is also used in some global optimisation algorithms based on set covering; see, for example, Zhigljavsky (1991). The idea is as follows. Assume that $f(\cdot)$ satisfies a Lipschitz condition of the type

$$|f(x) - f(x')| \leq \rho\|x - x'\| \quad \forall (x, x') \in \mathcal{R}_0 \times \mathcal{R}_0.$$

Then, having evaluated $f(\cdot)$ at $\boldsymbol{x}^{(1)}, \ldots, \boldsymbol{x}^{(k)}$, with

$$f(\hat{\boldsymbol{x}}^{(k)}) = \min_{n=1,\ldots,k} f(\boldsymbol{x}^{(n)}),$$

we know that $\boldsymbol{x}^*$ does not belong to $\cup_{n=1}^{k} \mathcal{B}(\boldsymbol{x}^{(n)}, \rho^{(n)})$, with

$$\rho^{(n)} = [f(\boldsymbol{x}^{(n)}) - f(\hat{\boldsymbol{x}}^{(k)})]/\rho.$$

The consistency set for $\boldsymbol{x}^*$ is thus

$$\mathcal{C}_k = \mathcal{R}_0 \setminus \cup_{n=1}^{k} \mathcal{B}(\boldsymbol{x}^{(n)}, \rho^{(n)}).$$

The points $\boldsymbol{x}^{(n)}$ can be generated sequentially or in bunches. A possible sequential approach, called *active covering*, is to select $\boldsymbol{x}^{(k+1)}$ as the minimiser of the Lipschitzian minorant $f^{(k)}(\cdot)$ of $f(\cdot)$,

$$f^{(k)}(\boldsymbol{x}) = \max_{n=1,\ldots,k} (f(\boldsymbol{x}^{(n)}) - \rho\|\boldsymbol{x} - \boldsymbol{x}^{(n)}\|);$$

see Pinter (1995), as shown in Figure 2.7. However the point $\boldsymbol{x}^{(k+1)}$ is difficult to determine when $d > 1$, since the graph of $f^{(k)}(\cdot)$ is obtained by intersecting cones in $\mathbb{R}^{d+1}$; see, *e.g.*, Evtushenko (1971) and the survey paper Rinnooy-Kan and Timmer (1989). In non-sequential methods (*passive covering*) the sequence $(\boldsymbol{x}^{(n)})$ does not depend on the observed values of $f(\cdot)$ and can be generated beforehand. Random or quasi-random sequences can be used. Note that $f(\boldsymbol{x}^{(k)})$ need not be evaluated when the k-th point in the sequence does not belong to the current consistency set. One may refer to Zhigljavsky and Chekmasov (1996) for a comparison between different covering schemes. In practise the Lipschitz constant can be estimated in the course of optimisation; see Zhigljavsky (1991). In this case, the consistency sets are not necessarily imbedded.

2.3.4 Ellipsoid algorithms

Outer ellipsoids for linear programming

Consider first the linear-programming problem of Section 2.3.1, and take an ellipsoid

$$\mathcal{E}_0 = \mathcal{E}(\boldsymbol{x}^{(0)}, \boldsymbol{M}_0) = \{\boldsymbol{x} \in \mathbb{R}^d \ / \ (\boldsymbol{x} - \boldsymbol{x}^{(0)})^T \boldsymbol{M}_0^{-1} (\boldsymbol{x} - \boldsymbol{x}^{(0)}) \leq 1\}$$

large enough to contain $\boldsymbol{x}^*$. Since $\mathbf{c}^T\boldsymbol{x}^* \leq \mathbf{c}^T\boldsymbol{x}^{(0)}$, $\boldsymbol{x}^*$ belongs to the half-ellipsoid defined by

$$\mathcal{E}_0 \cap \{\boldsymbol{x} \in \mathbb{R}^d \ / \ \mathbf{c}^T\boldsymbol{x} \leq \mathbf{c}^T\boldsymbol{x}^{(0)}\}.$$

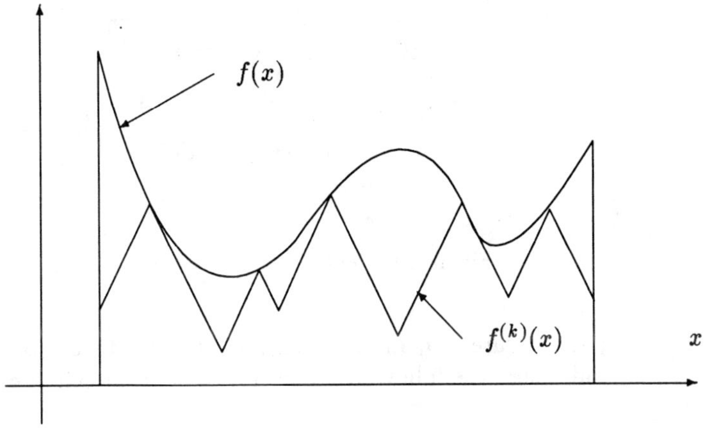

Figure 2.7 $f(\cdot)$ *and the minorant function* $f^{(k)}(\cdot)$

We can then construct the minimum-volume ellipsoid $\mathcal{E}_1$ that contains this half-ellipsoid (and thus necessarily $\boldsymbol{x}^*$), and repeat the operations. A key point is that at each iteration k the ellipsoid $\mathcal{E}_{k+1} = \mathcal{E}(\boldsymbol{x}^{(k+1)}, \boldsymbol{M}_{k+1})$ is easily obtained as follows; see, *e.g.*, Bland, Goldfarb and Todd (1981):

$$\left\{ \begin{array}{lcl} \boldsymbol{x}^{(k+1)} & = & \boldsymbol{x}^{(k)} - \rho \dfrac{\boldsymbol{M}_k \mathbf{c}}{\sqrt{\mathbf{c}^T \boldsymbol{M}_k \mathbf{c}}}, \\ \boldsymbol{M}_{k+1} & = & \sigma \left(\boldsymbol{M}_k - \tau \dfrac{\boldsymbol{M}_k \mathbf{c}\mathbf{c}^T \boldsymbol{M}_k}{\mathbf{c}^T \boldsymbol{M}_k \mathbf{c}} \right), \end{array} \right. \tag{2.22}$$

with

$$\rho = \frac{1}{d+1}, \quad \sigma = \frac{d^2}{d^2-1}, \quad \tau = \frac{2}{d+1}.$$

At some point, however, $\boldsymbol{x}^{(k)}$ may violate some of the constraints $\boldsymbol{A}\boldsymbol{x} \leq \boldsymbol{b}$. Consider then one violated constraint (for example, the most violated), that is $\boldsymbol{a}_i^T \boldsymbol{x}^{(k)} > b_i$, with $\boldsymbol{a}_i$ and b_i respectively the i-th row of $\boldsymbol{A}$ and the i-th component of $\boldsymbol{b}$. Two variants may then be used.

In the *central-cut* method, we simply substitute $\boldsymbol{a}_i$ for $\mathbf{c}$ in (2.22): the cut is still through the centre of $\mathcal{E}_k$, but oriented according to the violated constraint; see Figure 2.8.

In the *deep-cut* method, the cut is along the constraint, and the next ellipsoid has a smaller volume than with the central-cut

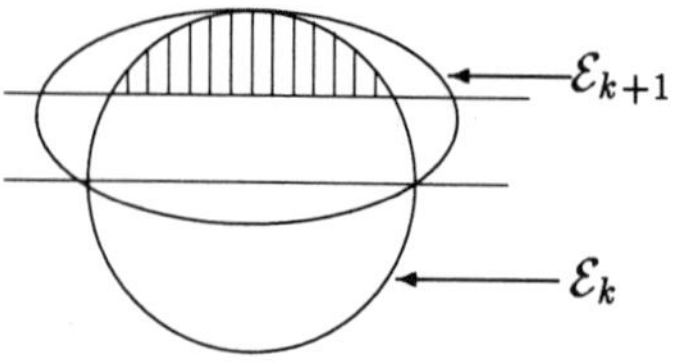

Figure 2.8 *Central-cut algorithm*

method; see Figure 2.9. In that case, we substitute $\boldsymbol{a}_i$ for $\mathbf{c}$ in (2.22), and assign the following values for the parameters ρ, σ and τ:

$$\rho = \frac{1 + d\alpha}{d + 1}, \quad \sigma = \frac{d^2(1 - \alpha^2)}{d^2 - 1}, \quad \tau = \frac{2(1 + d\alpha)}{(d + 1)(\alpha + 1)}, \tag{2.23}$$

where

$$\alpha = \alpha_k = \frac{\boldsymbol{a}_i^T \boldsymbol{x}^{(k)} - b_i}{\sqrt{\boldsymbol{a}_i^T \boldsymbol{M}_k \boldsymbol{a}_i}} > 0 .$$

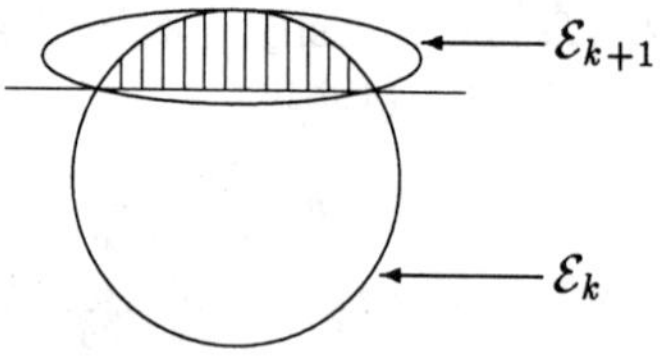

Figure 2.9 *Deep-cut algorithm*

Note that ρ, σ and τ vary with k in the deep-cut method. Also, the consistency set $\mathcal{C}_k$ defined by (2.17) is updated only when the cut is made by the objective. We may thus consider a subsequence (l_k) of (k) that counts the iterations of the algorithm where the cut is made by the objective, and $\mathcal{C}_{l_k} = \mathcal{C}_{l_{k+1}}$ if the cut is by the constraint. Then, $\mathcal{C}_{l_k} \subset \mathcal{E}_k$ implies $\mathcal{C}_{l_{k+1}} \subset \mathcal{E}_{k+1}$, $\forall k \geq 1$.

Constraints that are not violated by $\boldsymbol{x}^{(k)}$ could also be used to construct a new ellipsoid. The updating rule is then the same as for deep cuts, provided $\alpha > -1/d$ (when $\alpha \leq -1/d$, the minimum-volume ellipsoid containing the truncated ellipsoid is $\mathcal{E}_k$ itself).

Although this *shallow-cut* method might seem of no particular interest, it can be used to construct the maximum-volume inner ellipsoid for a polytope, as discussed in Section 6.1.4.

We now define the convergence rate at iteration k by

$$r_k = \frac{\mathtt{vol}(\mathcal{E}_{k+1})}{\mathtt{vol}(\mathcal{E}_k)} = \sigma^{d/2}\sqrt{1-\tau}\,,$$

which, for fixed d, is only a function of α. For the central-cut algorithm, $\alpha_k = 0$ for all k, and the rate is

$$r_k = r(d) = d^d\sqrt{1 - 2/(1+d)}/(d^2-1)^{d/2}\,,\ \forall k \geq 1\,. \qquad (2.24)$$

For $d = 2$ we get $r \simeq 0.7698$.

Outer ellipsoids for convex programming

Let $f(\cdot)$ be a convex function to be minimised, with $\boldsymbol{x}^* \in \mathcal{X} \subset I\!R^d$ the point where it achieves its minimum value. The constraints defining the feasible set $\mathcal{X}$ are assumed to be such that $\mathcal{X}$ is convex. Take an initial ellipsoid $\mathcal{E}_0$ containing $\boldsymbol{x}^*$. At iteration k, the cut is either through the centre $\boldsymbol{x}^{(k)}$ of $\mathcal{E}_k$ along the hyperplane tangent to the level set of the objective function, or along the tangent to a constraint violated by $\boldsymbol{x}^{(k)}$, if any (in this case, central cuts or deep cuts can be used). When the cut is central, the convergence rate is given by (2.24).

Figure 2.10 shows an iteration of the algorithm for $d = 2$ when there are no constraints and the objective function is $f(x_1, x_2) = (x_1-1)^2 + 4(x_2-1)^2$. The cut is through the centre of $\mathcal{E}_k$ along the tangent to the level set.

Although the convergence of the outer-ellipsoid algorithm is rather slow, this method has been used to prove that linear programming could be solved by algorithms with polynomial complexity (Khachiyan, 1979). A comparison with more classical optimisation algorithms for both convex and non-convex problems is presented in Ecker and Kupferschmid (1985). The (outer) ellipsoid algorithm is very efficient in the initial phase of the optimisation and relatively insensitive to a lack of precision in the evaluation of the objective function.

Inner ellipsoids

We first describe the method for the case of convex programming; see Tarasov, Khachiyan and Erlich (1988). Then we present an

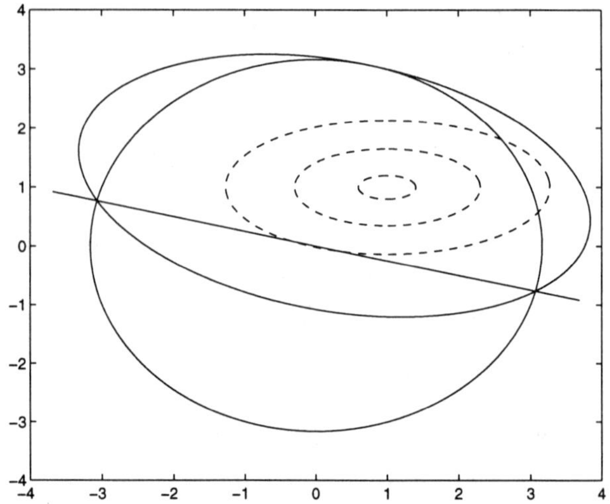

Figure 2.10 *One iteration of the outer-ellipsoid algorithm in convex programming: $\mathcal{E}_k$ and $\mathcal{E}_{k+1}$ are in full line; the contour levels of the objective function are in dashed lines*

alternative to be used for linear programming. The method of inner ellipsoids can be considered as a particular type of *interior-point* algorithm; see Wright (1998) for a survey.

The idea for convex programming is similar to that of cut-off methods; see Section 2.3.2. The point $\boldsymbol{x}^{(k)}$ is given by the centre of the maximum-volume ellipsoid $\mathcal{E}_i^*(\mathcal{C}_k)$ contained in $\mathcal{C}_k$. This ellipsoid is unique due to the Danzer-Zagustin Theorem; see Danzer, Grünbaum and Klee (1963). One can then show that, whatever the dimension d, the rate of convergence satisfies

$$r_k = \frac{\mathtt{vol}[\mathcal{E}_i^*(\mathcal{C}_{k+1})]}{\mathtt{vol}[\mathcal{E}_i^*(\mathcal{C}_k)]} \leq 0.843\,;$$

see Tarasov, Khachiyan and Erlich (1988). Since

$$\delta^n \leq \left(\frac{\mathtt{vol}[\mathcal{E}_i^*(\mathcal{C}_n)]}{\mathtt{vol}[\mathcal{E}_i^*(\mathcal{C}_0)]}\right)^{1/n},$$

where δ^n is defined by (2.18), this gives the bound

$$N(\epsilon) \leq 5.86\, d \log(1/\epsilon)$$

for the number of iterations required to obtain $\delta^n \leq \epsilon$. This result neglects the fact that inner ellipsoids with maximum volume cannot be determined exactly. It is shown in Tarasov, Khachiyan and Erlich (1988) that when using approximately optimal inner ellipsoids $\mathcal{E}^*_{i\,\gamma}(\mathcal{C}_k)$ such that $\mathtt{vol}[\mathcal{E}^*_{i\,\gamma}(\mathcal{C}_k)] \geq \gamma\ \mathtt{vol}[\mathcal{E}^*_i(\mathcal{C}_k)]$, then, for $\gamma = 0.99$, one gets the bound

$$N(\epsilon) \leq 6.64\, d \log(1/\epsilon)\,. \tag{2.25}$$

This value is worse than the bound (2.21) obtained for the method of centroids of Section 2.3.2 but, in contrast, each iteration of the algorithm now requires the solution of a problem of polynomial complexity. One can refer to Khachiyan and Todd (1993) and Pronzato and Walter (1993, 1996) for algorithms for the determination of maximum-volume ellipsoids contained in a polytope. See also Section 6.1.4.

We consider now the special case of linear programming, and show how the bound (2.25) can be improved. The problem is defined by (2.15–2.16). Let $\mathcal{C}_k$ be the consistency set obtained after the evaluation of $\boldsymbol{c}^T\boldsymbol{x}^{(1)}, \ldots, \boldsymbol{c}^T\boldsymbol{x}^{(k)}$. The algorithm alternates two steps: (i) construct the maximum-volume ellipsoid $\mathcal{E}^*_i(\mathcal{C}_k)$ contained in $\mathcal{C}_k$, $\mathcal{E}^*_i(\mathcal{C}_k) = \mathcal{E}(\boldsymbol{y}^{(k)}, \boldsymbol{M}_k)$, and (ii) determine the point

$$\boldsymbol{x}^{(k+1)} = \arg \min_{\boldsymbol{x} \in \mathcal{E}^*_i(\mathcal{C}_k)} \boldsymbol{c}^T\boldsymbol{x} = \boldsymbol{y}^{(k)} - \frac{\boldsymbol{M}_k\boldsymbol{c}}{\sqrt{\boldsymbol{c}^T\boldsymbol{M}_k\boldsymbol{c}}}\,,$$

which defines $\mathcal{C}_{k+1}$; see Figure 2.11.

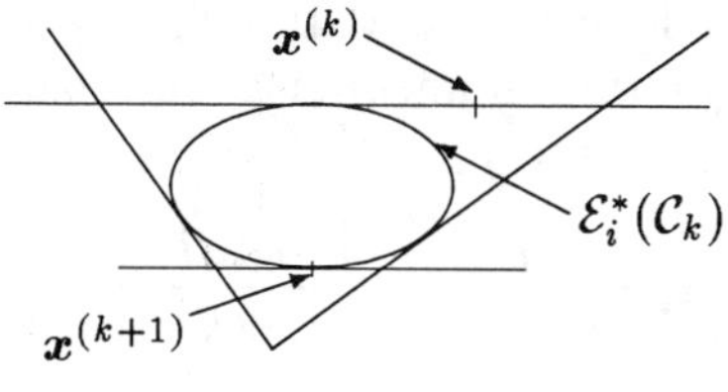

Figure 2.11 *Inner ellipsoids in linear programming*

We can easily obtain a bound for the rate of convergence of this method. Consider the ellipsoids $\mathcal{E}^*_i(\mathcal{C}_k)$ and $\mathcal{E}^*_i(\mathcal{C}_{k+1})$. As in Tarasov, Khachiyan and Erlich (1988), using appropriate transformation of coordinates we may assume that $\mathcal{E}^*_i(\mathcal{C}_k) = \mathcal{E}(\boldsymbol{a}, \boldsymbol{B}^2)$ and $\mathcal{E}^*_i(\mathcal{C}_{k+1}) = \mathcal{E}(-\boldsymbol{a}, \boldsymbol{B}^{-2})$, with $\boldsymbol{B} = \mathrm{diag}(b_1, \ldots, b_d)$ and $b_i > 0$, $i = 1, \ldots, d$. We then have $\|\boldsymbol{a}\| \geq 1$. Indeed, by construction there exists a hyperplane $\mathcal{H}$ that separates $\mathcal{E}_i^*(\mathcal{C}_k)$ and $\mathcal{E}_i^*(\mathcal{C}_{k+1})$; that is, there exists $\boldsymbol{u} \in \mathbb{R}^d$, with $||\boldsymbol{u}|| = 1$, such that

$$\forall \boldsymbol{x} \in \mathcal{E}_i^*(\mathcal{C}_k)\,,\ \boldsymbol{u}^T\boldsymbol{x} \leq 1 \text{ and } \forall \boldsymbol{x} \in \mathcal{E}_i^*(\mathcal{C}_{k+1})\,,\ \boldsymbol{u}^T\boldsymbol{x} \geq 1\,,$$

or

$$\forall \boldsymbol{x} \in \mathcal{E}_i^*(\mathcal{C}_{k+1})\,,\ \boldsymbol{u}^T\boldsymbol{x} \leq 1 \text{ and } \forall \boldsymbol{x} \in \mathcal{E}_i^*(\mathcal{C}_k)\,,\ \boldsymbol{u}^T\boldsymbol{x} \geq 1\,.$$

This gives $||\boldsymbol{B}\boldsymbol{u}|| + ||\boldsymbol{B}^{-1}\boldsymbol{u}|| \leq 2|\boldsymbol{u}^T\boldsymbol{a}|$, and thus

$$\begin{aligned} ||\boldsymbol{a}|| &\geq |\boldsymbol{u}^T\boldsymbol{a}| \geq \frac{1}{2}(||\boldsymbol{B}\boldsymbol{u}|| + ||\boldsymbol{B}^{-1}\boldsymbol{u}||) \\ &\geq ||\frac{\boldsymbol{B}+\boldsymbol{B}^{-1}}{2}\boldsymbol{u}|| \geq ||\boldsymbol{u}|| = 1\,. \end{aligned}$$

Now, $\mathcal{C}_k$ contains the origin as an interior point (since $\boldsymbol{a}$ and $-\boldsymbol{a}$ belong to $\mathcal{C}_k$) and is thus defined by a set of linear inequalities of the form $\boldsymbol{x}^T\boldsymbol{a}_i \leq 1$, $i = 1, \ldots, M$. Since $\mathcal{E}_i^*(\mathcal{C}_k) \subset \mathcal{C}_k$, one has $||\boldsymbol{B}\boldsymbol{a}_i|| \leq 1 - \boldsymbol{a}^T\boldsymbol{a}_i$, $i = 1, \ldots, M$. Similarly, $\mathcal{E}_i^*(\mathcal{C}_{k+1}) \subset \mathcal{C}_k$ implies $||\boldsymbol{B}^{-1}\boldsymbol{a}_i|| \leq 1 + \boldsymbol{a}^T\boldsymbol{a}_i$, $i = 1, \ldots, M$. This gives

$$||\boldsymbol{a}_i||^2 \leq ||\boldsymbol{B}\boldsymbol{a}_i||\ ||\boldsymbol{B}^{-1}\boldsymbol{a}_i|| \leq 1 - (\boldsymbol{a}^T\boldsymbol{a}_i)^2\,,\ i = 1, \ldots, M\,.$$

Thus, $\boldsymbol{a}_i$ belongs to the ellipsoid defined by $||\boldsymbol{x}||^2 + (\boldsymbol{a}^T\boldsymbol{x})^2 \leq 1$, obtained by contracting the unit ball by a factor $\sqrt{1+||\boldsymbol{a}||^2}$ in the direction of $\boldsymbol{a}$. Moreover, $\mathcal{C}_k$ contains the polar of this ellipsoid, that is, the ellipsoid obtained by expanding the unit ball by a factor $\sqrt{1+||\boldsymbol{a}||^2}$ in the direction of $\boldsymbol{a}$. Since $\mathcal{E}_i^*(\mathcal{C}_k)$ is of maximum volume, it follows that $\texttt{vol}[\mathcal{E}_i^*(\mathcal{C}_k)] \geq \vartheta_d\sqrt{1+||\boldsymbol{a}||^2}$, with ϑ_d the volume of the d-dimensional unit ball. Since $||\boldsymbol{a}|| \geq 1$, $\texttt{vol}[\mathcal{E}_i^*(\mathcal{C}_k)] \geq \vartheta_d\sqrt{2}$. Finally,

$$\begin{aligned} \frac{\texttt{vol}[\mathcal{E}_i^*(\mathcal{C}_{k+1})]}{\texttt{vol}[\mathcal{E}_i^*(\mathcal{C}_k)]} &= \frac{\vartheta_d \det \boldsymbol{B}^{-1}}{\vartheta_d \det \boldsymbol{B}} = \frac{1}{(\det \boldsymbol{B})^2} \\ &= \frac{\vartheta_d^2}{(\texttt{vol}[\mathcal{E}_i^*(\mathcal{C}_k)])^2} \leq \frac{1}{2}\,. \end{aligned}$$

We can also consider the more realistic situation where ellipsoids are determined approximately. As in Tarasov, Khachiyan and Erlich (1988), we say that an ellipsoid $\mathcal{E}_{i\ \gamma}^*(\mathcal{C}_k)$ is γ-optimal if

$$\mathcal{E}_{i\ \gamma}^*(\mathcal{C}_k) \subset \mathcal{C}_k \text{ and } \texttt{vol}[\mathcal{E}_{i\ \gamma}^*(\mathcal{C}_k)] \geq \gamma\ \texttt{vol}[\mathcal{E}_i^*(\mathcal{C}_k)]\,.$$

Such γ-optimal ellipsoids can be determined in a finite number of iterations; see Khachiyan and Todd (1993). We then obtain as before

$$\frac{\mathtt{vol}[\mathcal{E}_i^*(\mathcal{C}_{k+1})]}{\mathtt{vol}[\mathcal{E}_{i\ \gamma}^*(\mathcal{C}_k)]} = \frac{\vartheta_d^2}{(\mathtt{vol}[\mathcal{E}_{i\ \gamma}^*(\mathcal{C}_k)])^2} \leq \frac{1}{2\gamma^2},$$

and thus

$$\mathtt{vol}[\mathcal{E}_i^*(\mathcal{C}_{k+1})] \leq \frac{\gamma^{-2}}{2}\,\mathtt{vol}[\mathcal{E}_{i\ \gamma}^*(\mathcal{C}_k)] \leq \frac{\gamma^{-2}}{2}\,\mathtt{vol}[\mathcal{E}_i^*(\mathcal{C}_k)].$$

This gives the bound

$$N(\epsilon) \leq \frac{d\log(\frac{1}{\epsilon})}{\log 2 + 2\log\gamma}$$

for the number of iterations required to achieve the precision ϵ for δ^n, given by (2.18). For $\gamma = 0.99$, we obtain

$$N(\epsilon) \leq 1.49\, d\log(1/\epsilon),$$

to be compared to the bound (2.25) obtained for general convex programming problems.

2.3.5 Convex programming using function comparisons

It is potentially useful to generalise the line search algorithms of Section 2.2, which only require function value comparison, to higher dimensions. The following, somewhat speculative section, discusses the implication of assuming only convexity.

Consider a quasi-convex function $f(\cdot)$, such that its level sets $\mathcal{L}_k = \{\boldsymbol{x} \in \mathbb{R}^d \ / \ f(\boldsymbol{x}) \leq f(\boldsymbol{x}^{(k)})\}$ are convex sets in $\mathbb{R}^d$. Assume that $f(\cdot)$ has been evaluated at $d+1$ points $\boldsymbol{x}^{(1)}, \ldots, \boldsymbol{x}^{(d+1)}$ that form the vertices of a non-degenerate d-dimensional simplex $\mathcal{S}$. We assume that these points are ordered with respect to values of $f(\cdot)$: $f(\boldsymbol{x}^{(1)}) \leq \cdots \leq f(\boldsymbol{x}^{(d+1)})$. It is clear that knowing the order is equivalent to all pairwise function comparisons. Assume, moreover, that $f(\boldsymbol{x}^{(d+1)}) > f(\boldsymbol{x}^{(d)})$. Then, the point $\boldsymbol{x}^*$ where $f(\cdot)$ is minimum cannot belong to the polyhedral cone defined by the vertex $\boldsymbol{x}^{(d+1)}$ and the d hyperplanes containing the faces of $\mathcal{S}$ to which $\boldsymbol{x}^{(d+1)}$ belongs; see Figure 2.12 where $\boldsymbol{x}^*$ cannot belong to the dashed area. Indeed, assume that this is the case. Then the quasi-convexity of $f(\cdot)$ implies that the level set defined by $f(\boldsymbol{x}^{(d)})$ contains the simplex defined by the vertices $\boldsymbol{x}^*, \boldsymbol{x}^{(1)}, \ldots, \boldsymbol{x}^{(d)}$. This is impossible since $\boldsymbol{x}^{(d+1)}$ belongs to this simplex.

Having evaluated $f(\cdot)$ at a collection of k points $\boldsymbol{x}^{(i)}$, $i = 1, \ldots, k$

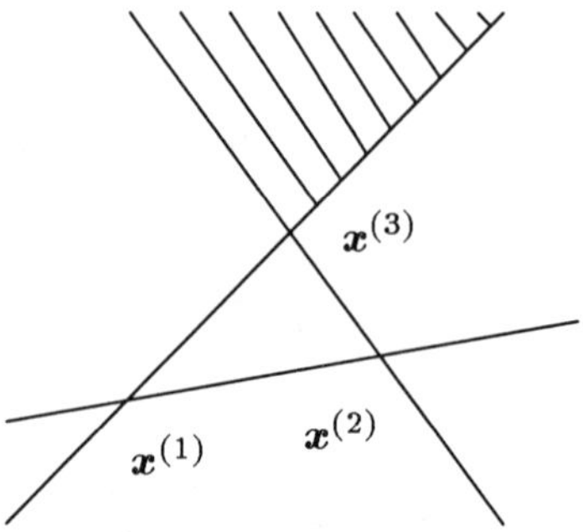

Figure 2.12 *Consistency for convex function using function comparisons*

$(k > d)$, the consistency set for x^* thus has the form

$$\mathcal{C}_k = \mathcal{R}_0 \setminus \cup\, \mathcal{C}^j \,,$$

where $\mathcal{C}^j$ are polyhedral cones obtained from the property above (see Figure 2.12). The function $f(\cdot)$ has already been evaluated at exactly d points in the interior of $\mathcal{C}_k$, and $\mathcal{C}_k$ is star shaped with respect to these points. A typical picture in dimension 2 is given in Figure 2.13.

Note that $\mathcal{C}_k$ may become exceedingly complicated as k increases, which raises the following possibilities for the choice of the sequence $(x^{(i)})$:

(i) the sequence should be such that the size of $\mathcal{C}_k$ (measured by its diameter, its volume, etc..) tends to zero as k increases,

(ii) the structure of the set $\mathcal{C}_k$ should remain as simple as possible (for example to allow an easy determination of its diameter or volume).

Note that choosing $x^{(k+1)}$ randomly in $\mathcal{C}_k$ ensures (i) but not (ii), and thus makes it difficult to generate subsequent points. A natural approach would then be to enclose $\mathcal{C}_k$ in a set $\mathcal{S}_k$ of simple shape, and to generate $x^{(k+1)}$ randomly in $\mathcal{S}_k$. At the same time, evaluations of $f(\cdot)$ out of $\mathcal{C}_k$ yield no information, in the sense that it does not produce any reduction of the size of $\mathcal{C}_k$. The approximating set $\mathcal{S}_k$ should thus be chosen as tight as possible, which is difficult due to the complicated star shape of $\mathcal{C}_k$.

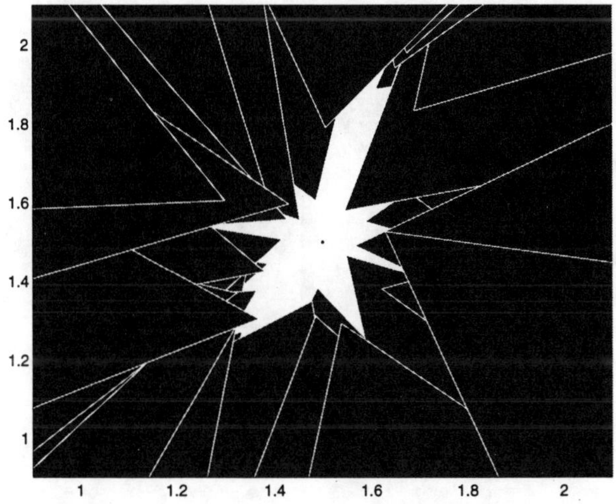

Figure 2.13 *Typical shape of the consistency region in dimension $d = 2$*

CHAPTER 3

Renormalisation

3.1 Towards a dynamical-system representation

3.1.1 General framework

As explained in Chapter 2, many search and optimisation algorithms for finding a target point $\boldsymbol{x}^* \in \mathcal{R}_0$ generate a sequence of regions $(\mathcal{R}_i)$, of the same shape, containing $\boldsymbol{x}^*$:

$$\mathcal{R}_1 \supseteq \mathcal{R}_0, \quad \mathcal{R}_1 \supseteq \mathcal{R}_2 \supseteq \ldots$$

Convergence to $\boldsymbol{x}^*$ takes place if the diameter of the sets converges to zero. The nesting of the regions occurs because additional observations or data are available at each stage creating additional assumptions on the set $\mathcal{R}_n$. Note that $\mathcal{R}_1$ should have a simple shape, and, as explained in Section 5.7.2, might be taken larger than $\mathcal{R}_0$.

Suppose that for each n there is a function $g_n(\cdot)$ mapping $\mathcal{R}_n$ back to some *base region* $\mathcal{R}$:

$$g_n : \mathcal{R}_n \longmapsto \mathcal{R}\,.$$

Let $\boldsymbol{x}_n = g_n(\boldsymbol{x}^*)$ be the location of the target in the renormalised region. The relation

$$\boldsymbol{x}_{n+1} = h_n(\boldsymbol{x}_n)$$

describes the path of the process for a given $\boldsymbol{x}^*$, with the starting value $\boldsymbol{x}_1 = g_1(\boldsymbol{x}^*)$. The mapping $h_n(\cdot)$ sets up a dynamical system $(\boldsymbol{x}_n)$ in $\mathcal{R}$ itself: instead of the target being fixed and $\mathcal{R}_n$ changing, $\mathcal{R}$ is fixed and the target moves. Figure 3.1 describes the mapping.

In this book the renormalisation function $g_n(\cdot)$ will typically be an *affine transformation*; see for example (3.9). However, in many situations, such as the line-search algorithms, the dynamical system cannot be defined in terms of $\boldsymbol{x}_n$ only. We shall denote by $T_n(\cdot)$ the updating rule for the state $\boldsymbol{z}_n$ of the dynamical system, which may depend on other variables. When necessary, we shall then distinguish the dimension of the search object $\boldsymbol{x}^*$ from the dimension

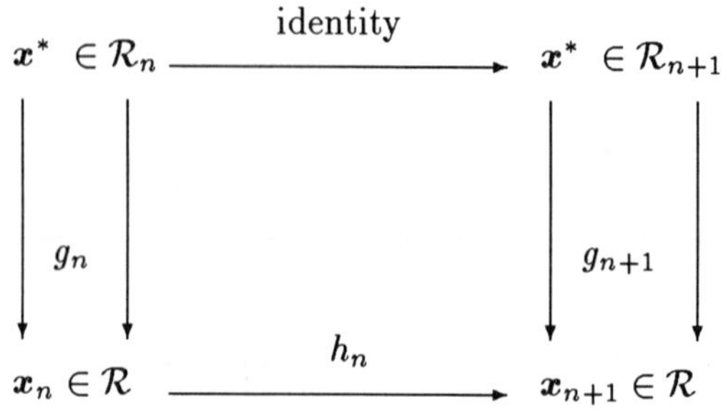

Figure 3.1 *Renormalisation*

of the state z_n. In particular, in optimisation problems, if the objective function $f(\cdot) = f_\theta(\cdot)$ belongs to a finite dimensional space with parameters θ, the parameters θ_n defining the renormalised function $f_n(\cdot)$ can be taken as state variables. The dynamical system then becomes time homogeneous, and z_n can be identified with (x_n, θ_n); see, for example, the ellipsoid algorithm of Section 3.3.1. Note that the condition of finite dimensionality of $f(\cdot)$ is not necessary; see, for instance, the dynamical process induced by the Golden-Section algorithm (Section 3.2). Also note that there are usually many possible choices for the state variables, and this choice is crucial for the complexity of the associated dynamical system.

In several important cases the regions $\mathcal{R}_n$ are larger than the region in which x^* is known to lie, the true consistency region $\mathcal{C}_n$, which may be difficult to renormalise. This trades a lower efficiency in the localisation of x^* with the benefit of a simpler renormalisation rule, or updating rule for the original process. This is true for an important example in optimisation, namely, the outer ellipsoid algorithm of linear and convex programming of Sections 2.3.4, 3.3.1. The region $\mathcal{R}_n$ may also be included in $\mathcal{C}_n$, provided it is the largest possible within some class of regions; see, for instance, the inner ellipsoid algorithm of Sections 2.3.4, 3.3.3. In this case, $\mathcal{R}_n$ does not necessarily contain x^*.

3.1.2 Dynamical-system representation

Consider the mapping $h_n(\cdot)$ of the last section. By tracing the diagram in Figure 3.1 in the clockwise direction we see that

$$h_n(\cdot) = g_{n+1}(g_n^{-1}(\cdot)) \,. \tag{3.1}$$

It is clear then that since $\mathcal{R}_n \supseteq \mathcal{R}_{n+1}$, the function $h_n(\cdot)$ is an expansion in the base region $\mathcal{R}$, which reflects the contraction in size from $\mathcal{R}_n$ to $\mathcal{R}_{n+1}$. This expansion is related to the rate of convergence of the algorithm. This relationship between the convergence rate of the algorithm and the rate of expansion of the associated dynamical system is at the centre of this work and is discussed in detail in Chapter 4.

In many cases, the form of the function $f(\cdot)$ is such that the dynamical system (z_n) defined in the previous section is *time homogeneous*; that is, the mapping $z_n \mapsto z_{n+1} = T(z_n)$ does not depend on n. Dynamical systems theory can then be used to study the ergodic behaviour of the search algorithm.

3.2 Renormalisation in line-search algorithms

3.2.1 First-order line search: root finding

Consider the bifurcation algorithm of Section 2.2.1, and assume that the function $f(\cdot)$ is linear. Let x_n denote the location of x^* in the current renormalised interval $[A_n, B_n)$,

$$x_n = \frac{x^* - A_n}{L_n} \, ;$$

then the updating rule for (x_n) is

$$x_{n+1} = T(x_n) = \begin{cases} 2x_n & \text{if } \ 0 \le x_n < \frac{1}{2} \\ 2x_n - 1 & \text{if } \ \frac{1}{2} \le x_n < 1 \end{cases}$$

This mapping, presented in Figure 3.2, is sometimes called the *Bernouilli shift*. It is well known to have an invariant measure which is uniform in $[0, 1)$. For starting values $x_1 = x^*$ which are *normal numbers** to the base 2, the asymptotic density of the sequence (x_n) is uniform on $[0, 1)$.

* A number in $[0, 1]$ is normal to the base 2 if the infinite sequence of 0's and 1's corresponding to its binary representation is such that each finite string of 0's and 1's of given length occurs with the same frequency in the sequence.

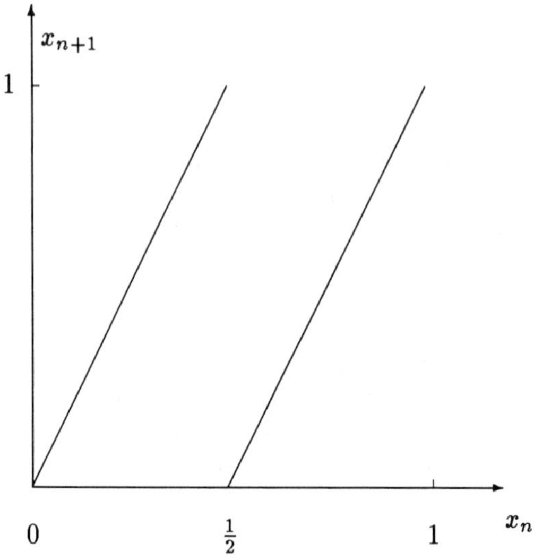

Figure 3.2 *Mapping of the Bernouilli shift*

Consider now the family of algorithms defined by (2.3). After right or left deletion we renormalise the uncertainty interval $[A_n, B_n)$ to $[0, 1)$. Assume, with no loss of generality, that $f(A_0) < f(B_0)$. Define e_n as the renormalised value of E_n, that is

$$e_n = \frac{E_n - A_n}{L_n}.$$

Then the algorithm satisfies $e_n = e$ if $|f_n(0)| \leq |f_n(1)|$ and $e_n = 1 - e$ otherwise, where $f_n(\cdot)$ is the renormalised objective function. Consider the case where $f(\cdot)$ is linear, which will be true asymptotically when $f(\cdot)$ is smooth at x^*. Then the condition $|f_n(0)| < |f_n(1)|$ is equivalent to $x_n < 1/2$, with x_n the renormalised value of x^*. The dynamical system describing the evolution of x_n is then

$$x_{n+1} = T(x_n) = \begin{cases} \frac{x_n}{e} & \text{if } 0 \leq x_n < e \\ \frac{x_n - e}{1-e} & \text{if } e \leq x_n < \frac{1}{2} \\ \frac{x_n}{1-e} & \text{if } \frac{1}{2} \leq x_n < 1 - e \\ \frac{x_n - (1-e)}{e} & \text{if } 1 - e \leq x_n < 1 \end{cases} \tag{3.2}$$

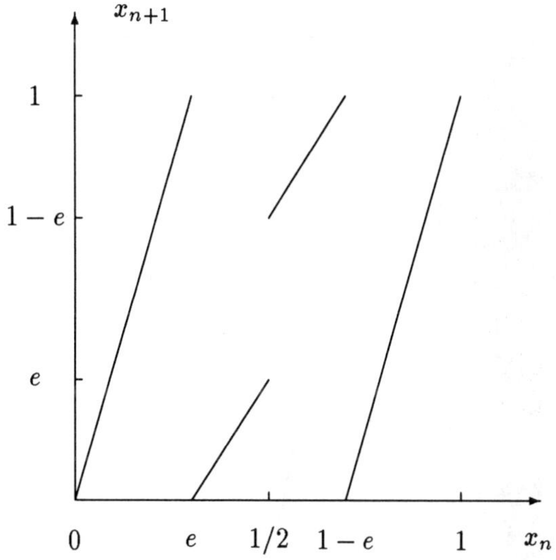

Figure 3.3 *Mapping of dynamical system (3.2)*

The mapping is shown in Figure 3.3 in the case $e = 1 - \sqrt{2}/2$.

We shall see in Section 5.1 that the dynamical systems associated with this family of line-search algorithms have invariant measures absolutely continuous with respect to the Lebesgue measure.

3.2.2 Continued fraction expansion and the Gauss map

The iteration (2.5) for $\epsilon_{n+1}(x^*) = \epsilon_{n+1}$ in Section 2.2.2 corresponds to a well-known dynamical system called the Gauss map. Since $\epsilon_n \in [0, 1)$, we can consider the process as already renormalised. The dynamical system is $\epsilon_n \mapsto \epsilon_{n+1} = T(\epsilon_n) = \{1/\epsilon_n\}$; see Figure 3.4. Its invariant density was given by Gauss in his diary dated October 25, 1880, and takes the form:

$$\phi(x) = \frac{1}{(1+x)\log 2}, \quad x \in [0, 1). \tag{3.3}$$

This describes the asymptotic behaviour of $\epsilon_n(x^*)$ for almost all x^* in $[0, 1)$; see Section 8.2.

We are interested in the behaviour of x^* in a suitably renormalised interval based on the "optimistic" and "pessimistic" con-

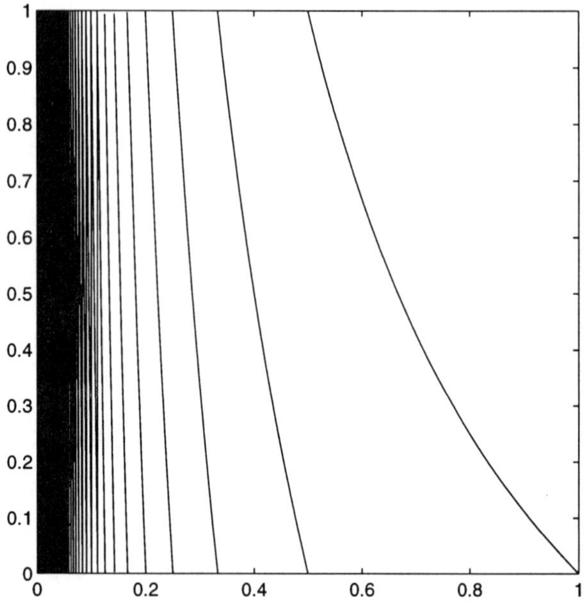

Figure 3.4 *The Gauss map*

sistency regions defined in Section 2.2.2, and in the lengths of these intervals.

For the pessimistic interval, define the length

$$L_n = \left| \frac{p_n}{q_n} - \frac{p_{n-1}}{q_{n-1}} \right| = \frac{1}{q_n q_{n-1}},$$

which is obtained from the identity (2.8). Then define the renormalised x^* as

$$x_{n+1} = \frac{\left| x^* - \frac{p_n}{q_n} \right|}{L_n}. \tag{3.4}$$

Since

$$x^* = \frac{p_n + \epsilon_{n+1} p_{n-1}}{q_n + \epsilon_{n+1} q_{n-1}}, \tag{3.5}$$

see (2.6), and using (2.8) again, we get

$$x_{n+1} = \frac{\epsilon_{n+1}}{\epsilon_{n+1} + \frac{q_n}{q_{n-1}}}.$$

Let $s_{n+1} = q_{n-1}/q_n$, which may be updated by using the identity

(2.7):

$$\frac{q_n}{q_{n-1}} = a_n + \frac{q_{n-2}}{q_{n-1}},$$

giving

$$s_{n+1} = \frac{1}{s_n + \left\lfloor \frac{1}{\epsilon_n} \right\rfloor},$$

where we used $a_n = \lfloor 1/\epsilon_n \rfloor$. Note that $s_1 = 0$ since $q_{-1} = 0$. Summarizing, we have

$$\left\{ \begin{array}{rcl} \epsilon_{n+1} & = & \frac{1}{\epsilon_n} - \left\lfloor \frac{1}{\epsilon_n} \right\rfloor \\ s_{n+1} & = & \frac{1}{s_n + \left\lfloor \frac{1}{\epsilon_n} \right\rfloor} \\ x_n & = & \frac{\epsilon_n s_n}{\epsilon_n s_n + 1} \end{array} \right. \tag{3.6}$$

with $\epsilon_1 = x^*$ and $s_1 = 0$.

The system for the optimistic interval uses the length

$$L_n = \left| \frac{p_n}{q_n} - \frac{p_n + p_{n-1}}{q_n + q_{n-1}} \right| = \frac{1}{q_n(q_{n-1} + q_n)}. \tag{3.7}$$

Using (3.5) and (2.8) again, we get the following process for x_n

$$x_n = \frac{\epsilon_n(1 + s_n)}{\epsilon_n s_n + 1},$$

the updating rules for ϵ_n and s_n being the same as previously.

Figure 3.5 (respectively 3.6) presents a plot of a typical sequence of iterates (x_n, s_{n+1}) in the pessimistic (respectively optimistic) case.

Continued fractions can be related to other dynamical systems in various ways, as a consequence of the relationship to the Gauss map. We give some classical examples.

The map

$$T_1(x) = \left\{ \begin{array}{ll} 1/x & \text{if } 0 < x \leq 1 \\ x - 1 & \text{if } 1 < x \end{array} \right.$$

is defined on $(0, \infty)$. Starting with $x^* \in (0, 1)$ and iterating, we immediately jump to the second branch and remain there until switching back to the first branch with the value $\{1/x^*\} = 1/x^* - \lfloor 1/x^* \rfloor$. This is repeated so that at the n-th switch back to the first branch we have the same ϵ_n as in the Gauss map. Moreover, the number of iterations spent in the second branch is precisely $a_n = \lfloor 1/\epsilon_n \rfloor$.

In the iteration just before coming to the first branch, the process

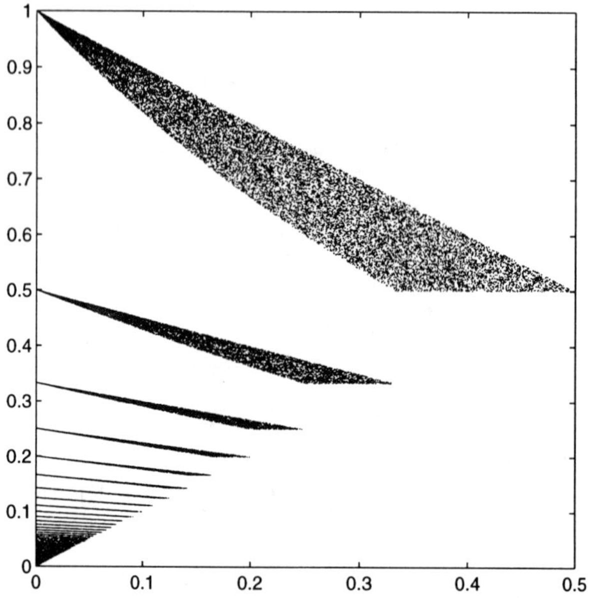

Figure 3.5 *Typical sequence of iterates* (x_n, s_{n+1}), $n = 1, \ldots, 50,000$, *in the pessimistic case*

is in $(1, 2]$. Consider a reduced mapping $T_2(\cdot)$, obtained by joining two iterations of $T_1(\cdot)$ when $x \in (1, 2]$. We have

$$T_2(x) = \begin{cases} T_1[T_1(x)] = 1/(x-1) & \text{if } 1 < x \leq 2 \\ T_1(x) = x - 1 & \text{if } 2 < x \end{cases}$$

Now, the change of variable $u = 1/x$ in $T_2(\cdot)$ gives the Farey map presented in Figure 3.7:

$$T_3(u) = \begin{cases} u/(1-u) & \text{if } 0 < u < 1/2 \\ (1-u)/u & \text{if } 1/2 \leq u < 1 \end{cases} \tag{3.8}$$

Note that the number of iterations spent in the second branch of $T_2(\cdot)$ before reaching the first, and in the first branch of $T_3(\cdot)$ before reaching the second, is $a_n - 1 = \lfloor 1/\epsilon_n \rfloor - 1$.

It is instructive to write the Frobenius-Perron equations for the three dynamical systems above and find the corresponding invariant densities (see Section 8.2 for definitions).

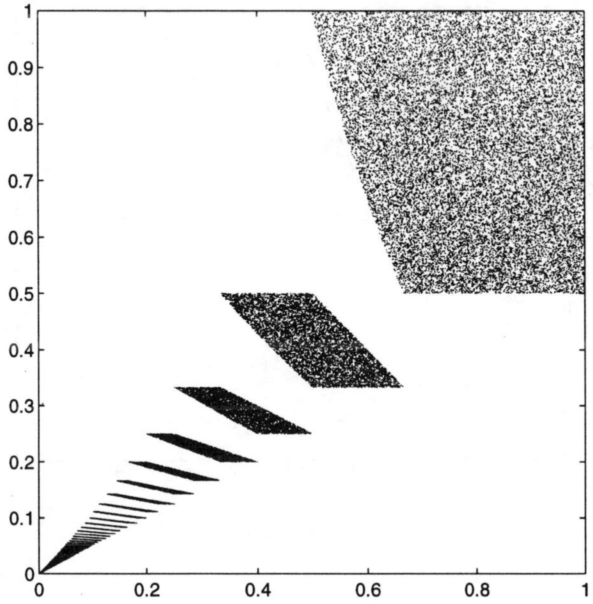

Figure 3.6 *Typical sequence of iterates* (x_n, s_{n+1}), $n = 1, \ldots, 50,000$, *in the optimistic case*

For the mapping $T_1(\cdot)$, the Frobenius-Perron equation is

$$\phi_1(y) = \begin{cases} \phi_1(y+1) & \text{if } y \leq 1 \\ \phi_1(y+1) + \frac{1}{y^2}\phi_1(\frac{1}{y}) & \text{if } 1 < y \end{cases}$$

and the invariant density, the solution of this equation, is

$$\phi_1(y) = \begin{cases} \frac{1}{y+1} & \text{if } y \leq 1 \\ \frac{1}{y} & \text{if } 1 < y \end{cases}$$

The density (3.3) of the Gauss map is achieved by truncating ϕ_1 to the interval $[0, 1]$ (with a normalisation). This is in complete agreement with the fact that the points in the sequence $(y_n = T_1^n(x^*))$ lying in $[0, 1]$ form the dynamical system $e_{n+1} = \{1/e_n\}$, $e_1 = x^*$.

For the mappings $T_2(\cdot)$ and $T_3(\cdot)$ the Frobenius-Perron equations and the invariant densities are respectively

$$\phi_2(y) \quad = \quad \phi_2(y+1) + \frac{1}{y^2}\phi_2(\frac{1}{y}+1),$$

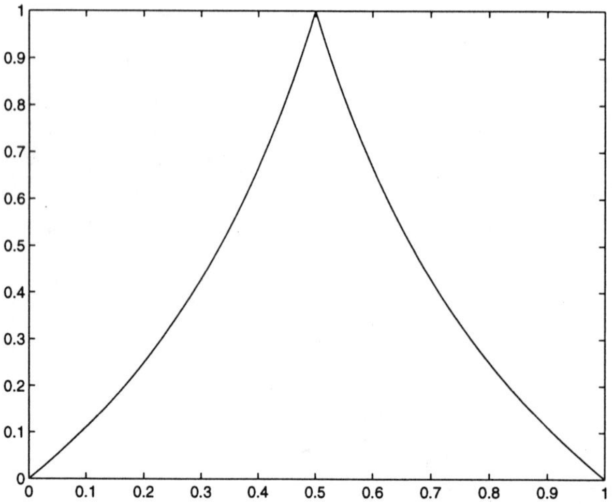

Figure 3.7 *The Farey map*

$$\phi_2(y) \quad = \quad \frac{1}{y}, \quad y > 1 \,,$$

and

$$\begin{aligned} \phi_3(y) &= \frac{1}{(y+1)^2}\left(\phi_3(\frac{y}{y+1}) + \phi_3(\frac{1}{y+1})\right), \\ \phi_3(y) &= \frac{1}{y}, \quad 0 < y < 1 \,. \end{aligned}$$

The dynamical systems $T_1(\cdot)$, $T_2(\cdot)$ and $T_3(\cdot)$ are expanding slower than the pure continued fraction expansion algorithm, because they typically undergo many iterations for one iteration of the continued fraction algorithm. This follows from the property of continued fractions that $\limsup_{n\to\infty} \mathbf{E}\{a_n(x^*)\} = \infty$ for almost all x^*, and is reflected in the fact that their invariant measures are not finite (they do not integrate). For the dynamical system $(x_{n+1} = T_3(x_n))$, this property is related to the fact that the slope of the mapping $T_3(\cdot)$ is unity at 0, a fixed point of the mapping, and causes the nonexponential divergence of the trajectories. Such a map is called *almost expanding* (see Section 8.2.4), which relates to the phenomenon called *chaos with intermittency*. This term is used to describe the occasional regular behaviour of the trajectories which occurs in the present case near 0.

3.2.3 Second-order line-search

Consider the line-search algorithms defined in Section 2.2.3. After left or right deletion, we renormalise each uncertainty interval $[A_n, B_n)$ to $[0, 1)$. Thus introduce normalised variables in $[0, 1)$

$$\begin{aligned} x_n &= g_n(x^*) = \tfrac{x^* - A_n}{L_n}\,, \\ e_n &= \tfrac{E_n - A_n}{L_n}\,,\quad e'_n = \tfrac{E'_n - A_n}{L_n}\,, \end{aligned} \tag{3.9}$$

and

$$u_n = \min(e_n, e'_n)\,,\ v_n = \max(e_n, e'_n)\,. \tag{3.10}$$

The deletion rule is

$$\begin{cases} \text{(R)}: & \text{if } f_n(u_n) < f_n(v_n) \text{ delete } [v_n, 1) \\ \text{(L)}: & \text{if } f_n(u_n) \geq f_n(v_n) \text{ delete } [0, u_n) \end{cases} \tag{3.11}$$

with $f_n(\cdot)$ the renormalised function on $[0, 1)$ defined by

$$f_n(x) = f[g_n^{-1}(x)]\,.$$

The remaining interval is then renormalised to $[0, 1)$. Straightforward calculation then shows that right and left deletions respectively give

$$x_{n+1} = \begin{cases} \frac{x_n}{v_n} & (R) \\ \frac{x_n - u_n}{1 - u_n} & (L) \end{cases}$$

Moreover, from the definition of E_{n+1}, we obtain

$$e_{n+1} = \begin{cases} \frac{u_n}{v_n} & (R) \\ \frac{v_n - u_n}{1 - u_n} & (L) \end{cases}$$

The GS algorithm corresponds to

$$v_n = 1 - u_n = \varphi\,,$$

which gives $x_{n+1} = h_n(x_n)$, with

$$h_n(x_n) = \begin{cases} x_n(1+\varphi) & \text{if } f_n(1-\varphi) < f_n(\varphi) \quad \text{(R)} \\ x_n(1+\varphi) - \varphi & \text{if } f_n(1-\varphi) \geq f_n(\varphi) \quad \text{(L)} \end{cases} \tag{3.12}$$

Assume that the function $f(\cdot)$ is symmetric with respect to x^*. Then the decision concerning left or right deletion at iteration n only depends on the position of x^* with respect to $(E_n + E'_n)/2$. In the renormalised form we thus obtain

$$\begin{cases} (R) & \text{if } x_n < \frac{e_n + e'_n}{2} \\ (L) & \text{if } x_n \geq \frac{e_n + e'_n}{2} \end{cases} \tag{3.13}$$

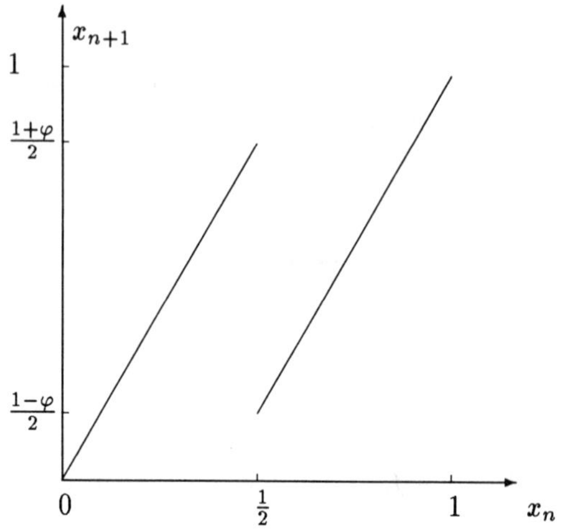

Figure 3.8 *The Golden-Section iteration*

For the GS algorithm, the updating rule (3.12) simply becomes

$$x_{n+1} = \begin{cases} (1+\varphi)x_n & \text{if } x_n < \frac{1}{2} \\ (1+\varphi)x_n - \varphi & \text{if } x_n \geq \frac{1}{2} \end{cases} \tag{3.14}$$

which defines a dynamical system; see Figure 3.8. Its ergodic behaviour is studied in Section 5.3. It is shown there that this ergodic behaviour is the same when $f(\cdot)$ is only locally symmetric with respect to x^*.

3.3 Renormalisation of ellipsoid algorithms

3.3.1 Outer ellipsoids for linear programming

We consider the ellipsoid algorithm of Section 2.3.4 in linear programming. As noted in Section 2.3.4, if the initial consistency set is included in $\mathcal{E}_1$, then the current consistency set is included in $\mathcal{R}_k = \mathcal{E}(\boldsymbol{x}^{(k)}, \boldsymbol{M}_k)$ for all $k \geq 1$, with

$$\mathcal{E}(\boldsymbol{x}^{(k)}, \boldsymbol{M}_k) = \{\boldsymbol{x} \in I\!R^d \;/\; (\boldsymbol{x} - \boldsymbol{x}^{(k)})^T \boldsymbol{M}_k^{-1} (\boldsymbol{x} - \boldsymbol{x}^{(k)}) \leq 1\}\,.$$

It is straightforward to renormalise an ellipsoid to the unit ball $\mathcal{B}(\boldsymbol{0}, 1)$, which is our base region $\mathcal{R}$. Additionally, the figure is rotated so that the objective is perpendicular to the d-th axis with coordinate vector $\boldsymbol{e}_d = (0, \ldots, 0, 1)^T$.

We specialize to the case of exactly d linear constraints of the form $\boldsymbol{Ax} \leq \boldsymbol{b}$, with the objective $\boldsymbol{c}^T\boldsymbol{x}$ to be minimised; see Figure 2.5. Let $\boldsymbol{P}_k$ be such that $\boldsymbol{P}_k\boldsymbol{P}_k^T = \boldsymbol{M}_k$ for all $k \geq 0$. First renormalise $\mathcal{R}_0$ to $\mathcal{R}$, which gives in the new coordinates

$$\boldsymbol{A}_1' = \boldsymbol{AP}_0\,,\ \boldsymbol{b}_1' = \boldsymbol{b} - \boldsymbol{Ax}^{(0)}\,,\ \boldsymbol{c}_1' = \boldsymbol{P}_0\boldsymbol{c}\,.$$

Then choose $\boldsymbol{Q}_0$ as a rotation matrix such that $\boldsymbol{Q}_0^T\boldsymbol{Q}_0 = \boldsymbol{I}$ and $\boldsymbol{Q}_0\boldsymbol{P}_0\boldsymbol{c}_1' = \boldsymbol{e}_d$, which gives in the rotated system of coordinates

$$\boldsymbol{A}_1 = \boldsymbol{AP}_0\boldsymbol{Q}_0^T\,,\ \boldsymbol{b}_1 = \boldsymbol{b} - \boldsymbol{Ax}^{(0)}\,,\ \boldsymbol{c}_1 = \boldsymbol{e}_d\,.$$

We apply the standard ellipsoid algorithm in this new system of coordinates. Let $\mathcal{H}_k$ be the hyperplane that cuts $\mathcal{R}$ at iteration k,

$$\mathcal{H}_k = \{\boldsymbol{x} \in \mathbb{R}^d \ / \ \boldsymbol{s}_k^T\boldsymbol{x} = \beta_k\}\,.$$

When the cut is by the objective, $\boldsymbol{s}_k = \boldsymbol{e}_d$ and $\beta_k = 0$ (deeper cuts, with negative values of β_k, will be considered in Section 6.2.2). When the cut is by the i-th constraint, one should distinguish between the central and deep-cut versions. In the central-cut case, $\boldsymbol{s}_k^T = \boldsymbol{a}_{k_i}^T$, the i-th row of $\boldsymbol{A}_k$, and $\beta_k = 0$. In the deep-cut case, $\boldsymbol{s}_k^T = \boldsymbol{a}_{k_i}^T$ and $\beta_k = b_{k_i}$, the i-th component of $\boldsymbol{b}_k$.

The ellipsoid containing the consistent part of the sphere is $\mathcal{E}(\boldsymbol{y}_k, \boldsymbol{P}_k\boldsymbol{P}_k^T)$, with

$$\boldsymbol{y}_k = -\rho\boldsymbol{s}_k/||\boldsymbol{s}_k|| \tag{3.15}$$

and

$$\boldsymbol{P}_k = \sqrt{\sigma}\left(\boldsymbol{I} - (1 - \sqrt{1-\tau})\frac{\boldsymbol{s}_k\boldsymbol{s}_k^T}{||\boldsymbol{s}_k||^2}\right)\,, \tag{3.16}$$

where ρ, σ, τ are given by (2.23), with $\alpha = \alpha_k = -\beta_k/||\boldsymbol{s}_k||$ the algebraic distance from the origin to $\mathcal{H}_k$. The full iteration is then

$$\boldsymbol{A}_{k+1} = \boldsymbol{A}_k\boldsymbol{P}_k\boldsymbol{Q}_k^T\,,\ \boldsymbol{b}_{k+1} = \boldsymbol{b}_k - \boldsymbol{A}_k\boldsymbol{y}_k\,, \tag{3.17}$$

with $\boldsymbol{Q}_k$ a rotation matrix such that $\boldsymbol{Q}_k\boldsymbol{Q}_k^T = \boldsymbol{I}$ and $\boldsymbol{Q}_k\boldsymbol{P}_k\boldsymbol{e}_d = \boldsymbol{e}_d$. The dynamical system $(\boldsymbol{z}_k)$ is thus defined by the components of $\boldsymbol{A}_k$ and $\boldsymbol{b}_k$, and the transformation $\boldsymbol{z}_k \to \boldsymbol{z}_{k+1}$ is linear (up to the multiplication by $\boldsymbol{Q}_k$, which does not influence the rate of convergence). Figure 3.9 presents a typical sequence of iterates $(\boldsymbol{x}_k = \boldsymbol{A}_k^{-1}\boldsymbol{b}_k)$ in the region of renormalisation $\mathcal{B}(\boldsymbol{0}, 1)$ when $d = 2$ for the central-cut algorithm.

As in Section 2.3.4, we define the convergence rate at iteration

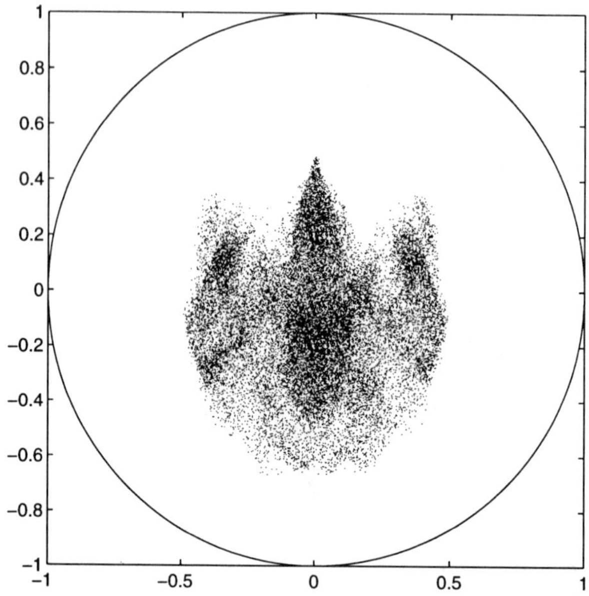

Figure 3.9 *Typical sequence of iterates for the central-cut algorithm in linear programming*

k by

$$r_k = \frac{\mathtt{vol}(\mathcal{E}_{k+1})}{\mathtt{vol}(\mathcal{E}_k)} = \mid \det J_k \mid = \sigma^{d/2}\sqrt{1-\tau}\,,$$

which, for fixed d, is function of α_k only. For the central-cut algorithm, $\alpha_k = 0$ for all k, and the rate is

$$r(d) = d^d\sqrt{1 - 2/(1+d)}/(d^2-1)^{d/2}\,,\ \forall k \geq 1\,, \tag{3.18}$$

which tends to 1 as d tends to infinity; see Figure 3.10. For $d = 2$, we obtain $r \simeq 0.7698$. In order to produce a fast rate of convergence (small r_k), α_k should be chosen as large as possible. However, simulations show that if the depth α_k of the constraint cut is fixed, the renormalised ellipsoid can be inconsistent with the data; that is, it may not contain $\boldsymbol{x}_k = \boldsymbol{A}_k^{-1}\boldsymbol{b}_k$. On the other hand, the depth α^0 for the objective cuts can be fixed to a positive value, while classical deep cuts $\alpha_k = -\beta_k/||\boldsymbol{s}_k||$ are used for the constraints; see Section 6.2.

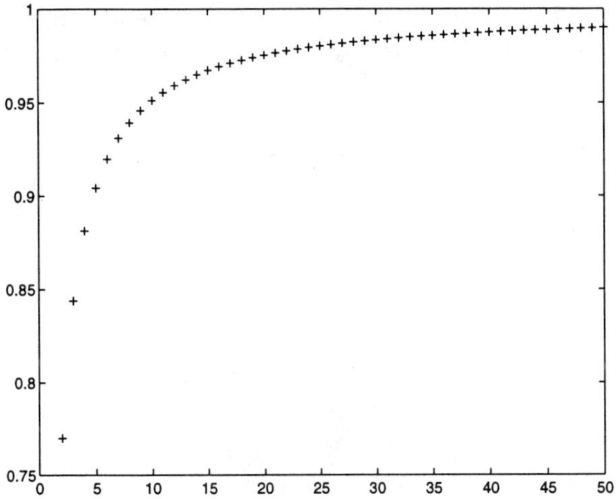

Figure 3.10 *Rate of convergence of the outer ellipsoid algorithm with central cuts as a function of d*

3.3.2 Outer ellipsoids for convex programming

Consider the algorithm of Section 2.3.4 for convex programming, in the case where the optimum is strictly feasible. The asymptotic behaviour of the algorithm is then as in the unconstrained case. We assume that the objective function to be minimised is quadratic,

$$f(\boldsymbol{x}) = \frac{1}{2}(\boldsymbol{x} - \boldsymbol{x}^*)^T \boldsymbol{A}(\boldsymbol{x} - \boldsymbol{x}^*) \,.$$

As in Section 3.3.1, we renormalise the current ellipsoid to the base region $\mathcal{R} = \mathcal{B}(\boldsymbol{0}, 1)$. Let $f_k(\boldsymbol{x})$ be the renormalised function,

$$f_k(\boldsymbol{x}) = \frac{1}{2}(\boldsymbol{x} - \boldsymbol{x}_k)^T \boldsymbol{A}_k(\boldsymbol{x} - \boldsymbol{x}_k) \,,$$

with $\boldsymbol{x}_k$ the renormalised value of $\boldsymbol{x}^*$ in $\mathcal{R}$. The updating rule is

$$\begin{aligned} \boldsymbol{x}_{k+1} &= \boldsymbol{P}_k^{-1}(\boldsymbol{x}_k - \boldsymbol{y}_k) \,, \\ \boldsymbol{A}_{k+1} &= \boldsymbol{P}_k^T \boldsymbol{A}_k \boldsymbol{P}_k \,, \end{aligned}$$

where $\boldsymbol{y}_k$ and $\boldsymbol{P}_k$ are given by one iteration of the ellipsoid algorithm initialised at $\mathcal{B}(\boldsymbol{0}, 1)$, with a cut through $\boldsymbol{0}$ orthogonal to the gradient $\boldsymbol{s}_k = -\boldsymbol{A}_k \boldsymbol{x}_k$; that is, $\boldsymbol{y}_k$ and $\boldsymbol{P}_k$ are given by (3.15, 3.16) with $\alpha = 0$. The convergence rate at each iteration is $r = \sigma^{d/2}\sqrt{1 - \tau}$, and $r \simeq 0.7698$ for $d = 2$.

Figure 3.11 presents a typical sequence of iterates $\boldsymbol{x}_k$ when, at each iteration, the figure is rotated so that the matrix $\boldsymbol{A}_k$ remains diagonal.

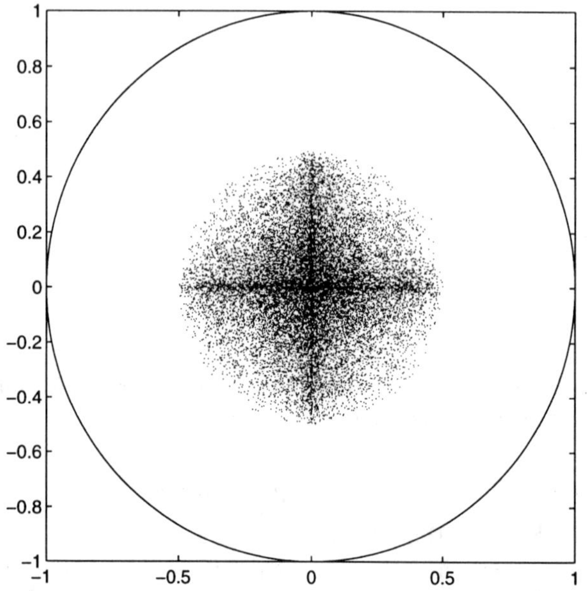

Figure 3.11 *Typical sequence of iterates for the central-cut algorithm in convex programming*

Numerical simulations indicate that one can cut deeper than through the centre of the current ellipsoid. Figure 3.12 shows the sequence of iterates (with rotation) when $\alpha = 0.2$. One gets in this case $r \simeq 0.6034$ for $d = 2$.

3.3.3 Inner ellipsoids

We consider the inner ellipsoid algorithm of Section 2.3.4 for linear programming. We specialize again to the case of exactly d linear constraints of the form $\boldsymbol{A}\boldsymbol{x} \leq \boldsymbol{b}$, with the objective $\boldsymbol{c}^T\boldsymbol{x}$ to be minimised. We renormalise $\mathcal{R}_0$ so that the initial search domain, after the evaluation of $\boldsymbol{c}^T\boldsymbol{x}^{(1)}$, forms a regular d-dimensional simplex; see Figure 3.13.

At any step k, the maximum-volume inner ellipsoid $\mathcal{E}_i^*(\mathcal{C}_k)$ is

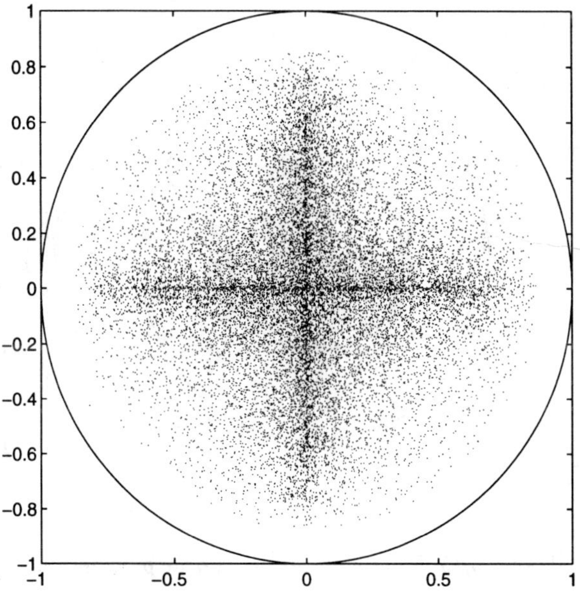

Figure 3.12 *Typical sequence of iterates for the outer-ellipsoid algorithm in convex programming with the depth of cuts fixed to 0.2*

then a sphere, and the ratio of volumes of two consecutive spheres is $[(d-1)/(d+1)]^d$ (this is due to the fact that the minimum-volume ellipsoid containing $\mathcal{C}_k$ is homothetic to $\mathcal{E}_i^*(\mathcal{C}_k)$ with a factor d; see Figure 3.13).

Since $f(\boldsymbol{x}) = \boldsymbol{c}^T\boldsymbol{x}$, one easily gets

$$\frac{\delta^{k+1}}{\delta^k} = \frac{d-1}{d+1},$$

where δ^k is given by (2.18), so that $\delta^k \leq (d-1)^k/(d+1)^k$, which gives the bound

$$N(\epsilon) \leq \frac{d-1}{2}\log(1/\epsilon)$$

for the number of iterations required to get $\delta^n \leq \epsilon$ (compare with the bounds of Section 2.3.4).

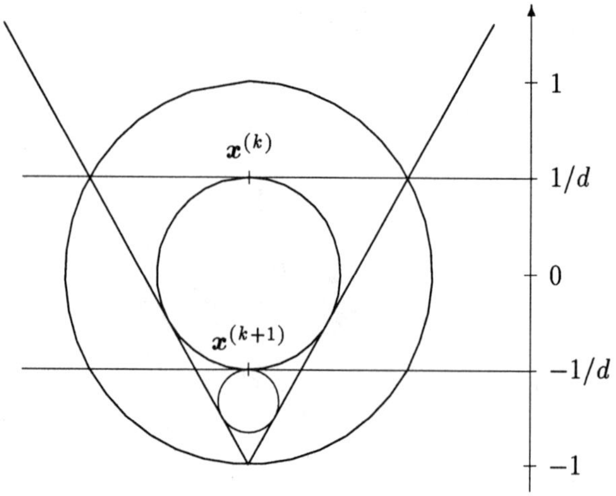

Figure 3.13 *Inner ellipsoids for linear programming*

3.4 Renormalisation in the steepest descent algorithm

Here we again normalise $\mathcal{R}_n$ to a unit sphere, but so that the optimising point $\boldsymbol{x}^*$ is fixed at the centre and its current approximation is on the boundary.

We assume that the objective function has the form

$$f(\boldsymbol{x}) = \frac{1}{2}(\boldsymbol{x} - \boldsymbol{x}^*)^T \boldsymbol{A}(\boldsymbol{x} - \boldsymbol{x}^*)$$

with $\boldsymbol{A}$ a positive-definite matrix. Without any loss of generality, we may assume that $\boldsymbol{A} = \mathrm{diag}(\lambda_1, \ldots, \lambda_d)$, where $0 < \lambda_1 \leq \ldots \leq \lambda_d$. Let $(\boldsymbol{x}^{(k)})$ denote the sequence of (unnormalised) iterates of the steepest descent algorithm. Then

$$\boldsymbol{x}^{(k+1)} = \boldsymbol{x}^{(k)} - \alpha_k \boldsymbol{A}(\boldsymbol{x}^{(k)} - \boldsymbol{x}^*),$$

where α_k is obtained by minimizing $[(\boldsymbol{I} - \alpha\boldsymbol{A})(\boldsymbol{x}^{(k)} - \boldsymbol{x}^*)]^T \boldsymbol{A}[(\boldsymbol{I} - \alpha\boldsymbol{A})(\boldsymbol{x}^{(k)} - \boldsymbol{x}^*)]$ with respect to α. This gives

$$\alpha_k = \frac{(\boldsymbol{x}^{(k)} - \boldsymbol{x}^*)^T \boldsymbol{A}^2(\boldsymbol{x}^{(k)} - \boldsymbol{x}^*)}{(\boldsymbol{x}^{(k)} - \boldsymbol{x}^*)^T \boldsymbol{A}^3(\boldsymbol{x}^{(k)} - \boldsymbol{x}^*)}$$

$$= \frac{\sum_{j=1}^{d}([\boldsymbol{x}^{(k)}]_j - [\boldsymbol{x}^*]_j)^2 \lambda_j^2}{\sum_{j=1}^{d}([\boldsymbol{x}^{(k)}]_j - [\boldsymbol{x}^*]_j)^2 \lambda_j^3},$$

to obtain

$$[\boldsymbol{x}^{(k+1)}]_i = [\boldsymbol{x}^{(k)}]_i - \alpha_k \lambda_i ([\boldsymbol{x}^{(k)}]_i - [\boldsymbol{x}^*]_i).$$

Define $\boldsymbol{z}_k$ by

$$\boldsymbol{z}_k = \frac{\boldsymbol{x}^{(k)} - \boldsymbol{x}^*}{||\boldsymbol{x}^{(k)} - \boldsymbol{x}^*||}.$$

We get

$$||\boldsymbol{x}^{(k+1)} - \boldsymbol{x}^*||^2 = \sum_{j=1}^{d} (1 - \alpha_k \lambda_j)^2 ([\boldsymbol{x}^{(k)}]_j - [\boldsymbol{x}^*]_j)^2,$$

which gives $\boldsymbol{z}_k \mapsto \boldsymbol{z}_{k+1} = T(\boldsymbol{z}_k)$, with

$$\begin{aligned} [\boldsymbol{z}_{k+1}]_i &= (1 - \alpha_k \lambda_i)[\boldsymbol{z}_k]_i \frac{||\boldsymbol{x}^{(k)} - \boldsymbol{x}^*||}{||\boldsymbol{x}^{(k+1)} - \boldsymbol{x}^*||} \\ &= \frac{(1 - \alpha_k \lambda_i)[\boldsymbol{z}_k]_i}{(\sum_{j=1}^{d} (1 - \alpha_k \lambda_j)^2 [\boldsymbol{z}_k]_i^2)^{1/2}}, \end{aligned}$$

and

$$\alpha_k = \frac{\sum_{j=1}^{d} \lambda_j^2 [\boldsymbol{z}_k]_j^2}{\sum_{j=1}^{d} \lambda_j^3 [\boldsymbol{z}_k]_j^2}.$$

Another renormalisation, leading to simpler developments, will be used in Chapter 7, where the behaviour of the steepest descent algorithm is considered in more detail.

3.5 Square algorithm

While the natural region for renormalisation in the ellipsoid algorithm is a sphere, other regions can be used, for example cubes, leading to a quite different behaviour. We consider the linear programming problem, and for the sake of simplicity restrict our attention to dimension $d = 2$.

3.5.1 Deep-cut algorithm

At step k, assume that the solution is known to lie in the unit square $\mathcal{R}_0 = \{\boldsymbol{x} \ / \ ||\boldsymbol{x}||_\infty \leq 1\}$ of the (x, y) plane. The orientation

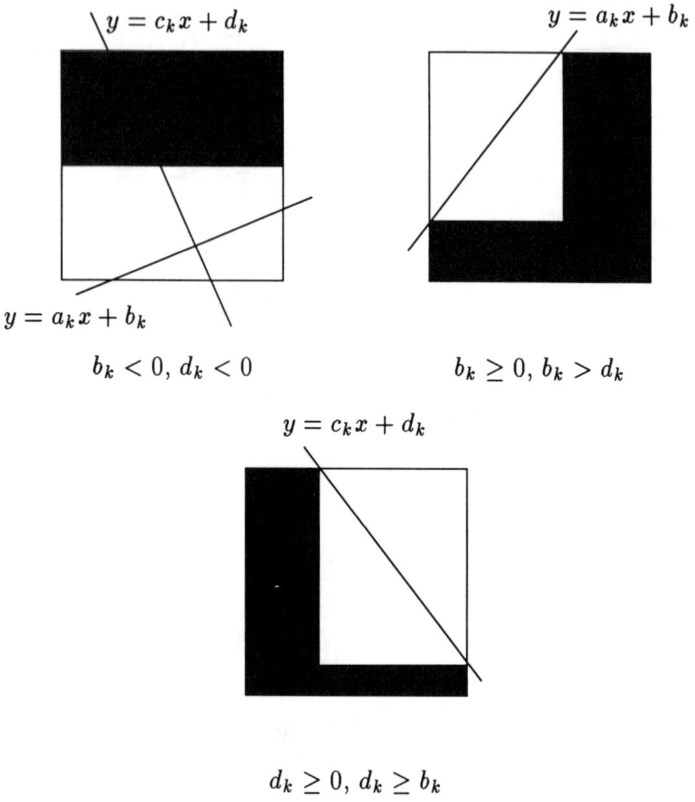

Figure 3.14 *The "square" algorithm with deep cuts for* $d = 2$

of the square is chosen such that the objective is $f(\boldsymbol{x}) = \boldsymbol{c}^T \boldsymbol{x}$, with $\boldsymbol{c} = (0\,,\ 1)^T$. The constraints are $y \geq a_k x + b_k$ and $y \geq c_k x + d_k$.

Consider central cuts for the objective case: when $b_k < 0$ and $d_k < 0$, cancel the upper half of the square, see Figure 3.14, and renormalise back to the unit square. Central cuts for the constraint case, parallel to the constraint and passing through the origin, do not yield a convergent algorithm. We thus consider deep cuts; that is, cancel the part under the most violated constraint at the origin, where the optimum cannot lie, see Figure 3.14, and renormalise back to the unit square.

It can easily be shown that the ratio c_k/a_k remains constant, and we shall denote its value by γ. When $\gamma = 1-2^k$ or $1/\gamma = 1-2^k$, the convergence of the unrenormalised process is finite, and we shall thus exclude this situation. The updating equations are then as follows, with r_k the reduction rate in terms of volume:

(i) if $b_k < 0\,,\ d_k < 0$:

$$a_{k+1} = 2a_k\,,\ b_{k+1} = 2b_k + 1\,,\ d_{k+1} = 2d_k + 1\,,\ r_k = \frac{1}{2}\,,$$

(ii) if $b_k \geq 0\,,\ b_k > d_k$:

$$\begin{aligned} a_{k+1} &= \tfrac{1+\alpha}{1-\beta}a_k\,, & b_{k+1} &= \tfrac{2b_k+a_k(\alpha-1)-(\beta+1)}{1-\beta}\,, \\ d_{k+1} &= \tfrac{2d_k+\gamma(\alpha-1)-(\beta+1)}{1-\beta}\,, & r_k &= \tfrac{(1+\alpha)(1-\beta)}{4}\,, \end{aligned}$$

where $\alpha = \min((1-b_k)/a_k, 1)$, $\beta = \max(b_k - a_k, -1)$,

(iii) if $d_k \geq 0\,,\ d_k \geq b_k$:

$$\begin{aligned} a_{k+1} &= \tfrac{1-\alpha}{1-\beta}a_k\,, & b_{k+1} &= \tfrac{2b_k+a_k(\alpha+1)-(\beta+1)}{1-\beta}\,, \\ d_{k+1} &= \tfrac{2d_k+\gamma a_k(\alpha+1)-(\beta+1)}{1-\beta}\,, & r_k &= \tfrac{(1-\alpha)(1-\beta)}{4}\,, \end{aligned}$$

where $\alpha = \max((1-d_k)/(\gamma a_k), -1)$, $\beta = \max(\gamma a_k + d_k, -1)$.

It can be shown that after a finite number of iterations the behaviour of the dynamical system

$$\boldsymbol{x}_k = (a_k, b_k, d_k) \mapsto \boldsymbol{x}_{k+1} = (a_{k+1}, b_{k+1}, d_{k+1})$$

is periodic. We assume first that $\gamma < -1$, define k_γ by

$$1 - 2^{k_\gamma+1} < \gamma < 1 - 2^{k_\gamma}\,,$$

and define the following states

$$\begin{aligned} S_j &= \begin{pmatrix} 1 \\ 0 \\ \gamma + 2^{j+1} - 1 \end{pmatrix}, \ j = 0, \ldots, k_\gamma\,, \\ S'_j &= \begin{pmatrix} 2 \\ 1 \\ 2\gamma + 2^{j+1} - 1 \end{pmatrix}, \ j = 1, \ldots, k_\gamma\,, \\ Z &= \begin{pmatrix} \frac{\gamma+2^{k_\gamma}-1}{\gamma} \\ \frac{1-2^{k_\gamma}}{\gamma} \\ \gamma + 2^{k_\gamma} \end{pmatrix}, \ Z' = \begin{pmatrix} -\frac{1}{\gamma} \\ \frac{\gamma+1}{\gamma} \\ 0 \end{pmatrix}. \end{aligned}$$

The behaviour of the process is then as follows:

$$S_j \xrightarrow{1/2} S'_{j+1} \xrightarrow{1/2} S_{j+1},\ j = 0, \ldots, k_\gamma - 1\,,$$

with reduction rate 1/2 for each transition. Then

(i) if $1 - 2^{k_\gamma+1} < \gamma < -2^{k_\gamma}$,

$$S_{k_\gamma} \xrightarrow{r_1} Z \xrightarrow{r_2} S_0\,,$$

with $r_1 = 1 - (1 - 2^{k_\gamma})/\gamma$, $r_2 = 1 + (2^{k_\gamma} - 1)/\gamma$;

(ii) if $-2^{k_\gamma} \leq \gamma < 1 - 2^{k_\gamma}$,

$$S_{k_\gamma} \xrightarrow{r'_1} Z' \xrightarrow{r'_2} S_0\,,$$

with $r'_1 = (1 - \gamma - 2^{k_\gamma})(1 - (1 - 2^{k_\gamma})/\gamma)$, $r_2 = -1/\gamma$.

The period is $2k_\gamma + 2$, and the log-rate, defined by (4.3), is

$$\varrho = \frac{k_\gamma}{k_\gamma + 1}\log 2 - \frac{1}{k_\gamma + 1}\log(1 - \frac{1 - 2^{k_\gamma}}{\gamma})\,.$$

The log-rate ϱ is infinite for $\gamma = 1 - 2^k$, which corresponds to finite convergence for the original unnormalised process. The same study is valid for $-1 < \gamma < 0$, and the same results hold with γ replaced by $1/\gamma$. We have $\inf_\gamma \varrho = (\log 3)/2$.

3.5.2 Double cuts

The reduction rate of the algorithm can be improved by considering deep cuts above and under the constraints, as indicated in Figure 3.15 for the constraint $y \geq a_k x + b_k$.

When $\gamma = 1-2^k$ or $1/\gamma = 1-2^k$ the convergence is finite, and we exclude this situation. For other values of γ, the behaviour is again periodic, but with a more complex structure than for ordinary deep cuts. Consider the case $\gamma < -1$. One can show that the process passes through states of the form $(1, 0, a)$, with $a < 0$. We then have the following transitions, with reduction rates indicated above the arrows:

$$\begin{pmatrix} 1 \\ 0 \\ a \end{pmatrix} \xrightarrow{1/2} \begin{pmatrix} 2 \\ 1 \\ f_1(a) + \gamma \end{pmatrix} \xrightarrow{1/2} \begin{pmatrix} 1 \\ 0 \\ f_1(a) \end{pmatrix} \xrightarrow{1/2} \cdots$$

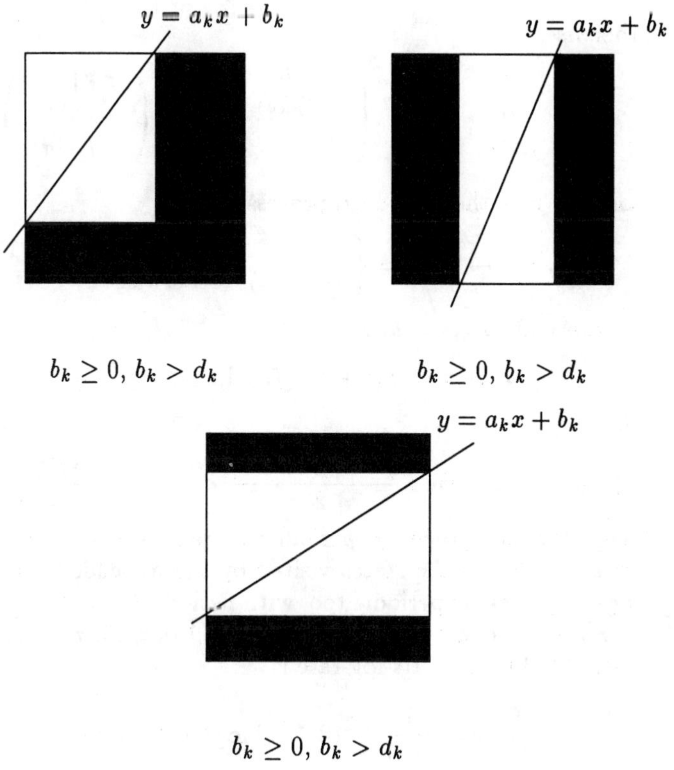

Figure 3.15 *The "square" algorithm with deep cuts above and under the constraints for* $d = 2$

$$\overset{1/2}{\longrightarrow} \begin{pmatrix} 2 \\ 1 \\ f_{j^*}(a) + \gamma \end{pmatrix} \overset{1/2}{\longrightarrow} \begin{pmatrix} 1 \\ 0 \\ f_{j^*}(a) \end{pmatrix},$$

with

$$f_j(a) = 2^j a + (2^j - 1)(1 - \gamma)$$

and $j^* = j^*(a)$ such that $f_{j^*-1}(a) \leq 0 < f_{j^*}(a)$. Then, if $f_{j^*}(a) > -(\gamma + 1)$

$$\begin{pmatrix} 1 \\ 0 \\ f_{j^*}(a) \end{pmatrix} \overset{-[1-\gamma-f_{j^*}(a)]/(4\gamma)}{\longrightarrow} \begin{pmatrix} -\frac{1}{\gamma} \\ \frac{\gamma+1}{\gamma} \\ 0 \end{pmatrix} \overset{-1/\gamma}{\longrightarrow} \begin{pmatrix} 1 \\ 0 \\ \gamma + 1 \end{pmatrix},$$

otherwise

$$\begin{pmatrix} 1 \\ 0 \\ f_{j^*}(a) \end{pmatrix} \overset{-1/\gamma}{\longrightarrow} \begin{pmatrix} -\frac{1}{\gamma} \\ -\frac{f_{j^*}(a)}{\gamma} \\ 0 \end{pmatrix} \overset{-1/\gamma}{\longrightarrow} \begin{pmatrix} 1 \\ 0 \\ -f_{j^*}(a) \end{pmatrix}.$$

Consider now the imbedded process

$$\omega_k = a \mapsto \omega_{k+1} = \begin{cases} \gamma + 1 & \text{if } f_{j^*}(a) > -(\gamma+1) \\ -f_{j^*}(a) & \text{otherwise}. \end{cases}$$

We can write $\omega_{k+1} = h(\omega_k)$, with

$$h(\omega) = \max[\gamma + 1, -f(\omega)], \ \gamma + 1 \leq \omega \leq 0,$$

and

$$f(\omega) = f_j(\omega) \text{ when } \frac{(1-\gamma)(1-2^j)}{2^j} < \omega \leq \frac{(1-\gamma)(1-2^{j-1})}{2^{j-1}}.$$

This imbedded process is periodic; its period n is a function of γ. Let $s_1, \ldots, s_n$ be the states visited by the imbedded process. The original process is periodic too, with period $\tau(\gamma) = \sum_{i=1}^{n} 2j^*(s_i) + 2$. One has for instance $\tau(-9) = 6$, $\tau(-10) = 16$, $\tau(-11) = 4$ and $\tau(-11.0001) = 102$. Its log-rate is

$$\varrho = \frac{1}{\tau(\gamma)} \sum_{i=1}^{n} (2j^*(s_i) \log 2 - 2 \log r_i),$$

with $r_i = [1-\gamma-f_{j^*}(s_i)]/(4\gamma^2)$ if $f_{j^*}(s_i) > -(\gamma+1)$ and $r_i = 1/\gamma^2$ otherwise. The worst log-rate is $\log 2$ obtained for $\gamma = -2$. The same study is valid for $-1 < \gamma < 0$, simply replacing γ by $1/\gamma$.

CHAPTER 4

Rates of convergence

4.1 Ergodic rates of convergence

Consider a one-dimensional (line-search) algorithm for an unknown target x^* in $\mathcal{R}_0 = [A_0, B_0)$, and define $\mathcal{R} = [0,1), \mathcal{R}_1 = [A_1, B_1) \supseteq \mathcal{R}_0$. The length L_n of the current uncertainty interval $\mathcal{R}_n = [A_n, B_n)$ containing x^* is $L_n = B_n - A_n$. This obviously depends on the location of x^* in $\mathcal{R}_0$, and thus on x_1, *i.e.*, $L_n = L_n(x_1)$. We define the reduction or *convergence rate* of the n-th iteration as

$$r_n = r_n(x_1) = \frac{L_{n+1}}{L_n} = \frac{B_{n+1} - A_{n+1}}{B_n - A_n}, \quad n \geq 0. \tag{4.1}$$

We also introduce

$$R = R(x_1) = \limsup_{N\to\infty}[L_N(x_1)]^{\frac{1}{N}} = \limsup_{N\to\infty}[L_0 \prod_{n=0}^{N-1} r_n(x_1)]^{\frac{1}{N}}. \tag{4.2}$$

If the limit exists and is the same for almost all x_1 with respect to the Lebesgue measure, then R will be called the *ergodic convergence rate*. Since $L_0 < \infty$ this becomes

$$R = \lim_{N\to\infty}(\prod_{n=1}^{N} r_n)^{\frac{1}{N}},$$

and the logarithmic form (called *log-rate*) is

$$\varrho = -\log R = -\lim_{N\to\infty}\frac{1}{N}\sum_{n=1}^{N}\log r_n \tag{4.3}$$

provided the limit exists for almost all x_1. The definition (4.1) of the convergence rate can be generalised to multidimensional algorithms, using the volumes of regions $\mathcal{R}_n$, *i.e.*, we define in this case

$$r_n = \frac{L_{n+1}}{L_n}, \quad n \geq 0, \tag{4.4}$$

where $L_n = \texttt{vol}(\mathcal{R}_n)$. We shall say that the algorithm has *exponential convergence* if $R(\boldsymbol{x}) < 1 - \alpha$ for all $\boldsymbol{x}$ and some $\alpha > 0$.

An important question concerns the relation between the log-rate ϱ and the *Lyapunov exponents* Λ_i, defined in Section 8.2.2. Consider the general case where $\boldsymbol{x}^* \in \mathbb{R}^d$. As already mentioned in Chapter 3, the renormalisation is typically *affine* with respect to $\boldsymbol{x}^*$; that is, the renormalised location of $\boldsymbol{x}^*$ in $\mathcal{R}_n$ satisfies

$$\boldsymbol{x}_n = g_n(\boldsymbol{x}^*) = \boldsymbol{\Omega}_n \boldsymbol{x}^* + \boldsymbol{\omega}_n \,,$$

where the non-singular square matrix $\boldsymbol{\Omega}_n$ and the "shift" vector $\boldsymbol{\omega}_n$ may depend on certain state variables $\boldsymbol{\theta}_n$ of the dynamical process. The variables $\boldsymbol{\theta}_n$ do not depend explicitly on $\boldsymbol{x}^*$ since they are related to known characteristics of the objective function or to parameters of the search algorithm. The full dynamical process is $\boldsymbol{z}_{n+1} = (\boldsymbol{x}_{n+1}, \boldsymbol{\theta}_{n+1}) = T(\boldsymbol{x}_n, \boldsymbol{\theta}_n)$, with, from (3.1),

$$\boldsymbol{x}_{n+1} = \boldsymbol{\Omega}_{n+1}\boldsymbol{\Omega}_n^{-1}\boldsymbol{x}_n - \boldsymbol{\Omega}_{n+1}\boldsymbol{\Omega}_n^{-1}\boldsymbol{\omega}_n + \boldsymbol{\omega}_{n+1} \,,$$

and $\boldsymbol{\theta}_{n+1}$ not depending explicitly on $\boldsymbol{x}_n$. Define $\boldsymbol{\Sigma}_n = \partial\boldsymbol{\theta}_{n+1}/\partial\boldsymbol{\theta}_n^T$. We then have the following theorem.

Theorem 4.1 *Assume that the matrices $\boldsymbol{\Omega}_n\boldsymbol{\Omega}_1^{-1}$ and $\prod_{i=1}^n \boldsymbol{\Sigma}_i$ have real eigenvalues for all n and that the Lyapunov exponents of the dynamical system $T(\cdot)$ are well defined (see Section 8.2.2); the log-rate ϱ is then equal to the sum of d Lyapunov exponents of $T(\cdot)$, with $d =$ dim $\boldsymbol{x}$.*

Proof. The Jacobian matrix of the transformation $T(\cdot)$ can be written in the block-triangular form

$$\boldsymbol{J}_T(\boldsymbol{z}_n) = \begin{pmatrix} \boldsymbol{\Omega}_{n+1}\boldsymbol{\Omega}_n^{-1} & \boldsymbol{\Sigma}'_n \\ 0 & \boldsymbol{\Sigma}_n \end{pmatrix},$$

where $\boldsymbol{\Sigma}'_n$ is some matrix. We thus get

$$\boldsymbol{J}_{T^n}(\boldsymbol{z}_1) = \boldsymbol{J}_T(\boldsymbol{z}_n) \times \cdots \times \boldsymbol{J}_T(\boldsymbol{z}_1) = \begin{pmatrix} \boldsymbol{\Omega}_{n+1}\boldsymbol{\Omega}_1^{-1} & \boldsymbol{\Sigma}''_n \\ 0 & \prod_{i=1}^n \boldsymbol{\Sigma}_i \end{pmatrix},$$

where $\boldsymbol{\Sigma}''_n$ is again some matrix. The Lyapunov exponents associated with the d components of $\boldsymbol{x}$ are

$$\Lambda_j = \lim_{n\to\infty} \frac{1}{n} \log \lambda_{j,n}(\boldsymbol{z}_1) \,,$$

where the $\lambda_{j,n}(\boldsymbol{z}_1)$ are the absolute values of the eigenvalues of

$\boldsymbol{\Omega}_{n+1}\boldsymbol{\Omega}_1^{-1}$, taken in decreasing order. We thus obtain

$$\sum_{j=1}^{d} \Lambda_j = \lim_{n\to\infty} \frac{1}{n} \log|\det \boldsymbol{\Omega}_{n+1}\boldsymbol{\Omega}_1^{-1}|.$$

Since $L_n = L_0/|\det \boldsymbol{\Omega}_n|$, we have

$$r_n = \frac{|\det \boldsymbol{\Omega}_n|}{|\det \boldsymbol{\Omega}_{n+1}|}, \quad n \geq 1,$$

and thus

$$\begin{aligned}\sum_{j=1}^{d} \Lambda_j &= \lim_{n\to\infty} \frac{1}{n} \log(\prod_{i=1}^{n} r_i^{-1}) \\ &= -\lim_{n\to\infty} \frac{1}{n} \sum_{i=1}^{n} \log r_i = \varrho.\end{aligned}$$

■

This theorem explains why, in the case $d = 1$, and in particular for the line-search algorithms of Section 2.2.3 and 2.2.3, ϱ is one of the Lyapunov exponents of the dynamical system, usually the largest. However, an example where $\boldsymbol{\theta}_n$ depends on x^*, and the theorem above thus does not apply, will be given in Section 5.2.

4.2 Characteristics of average performances

As already noticed, $L_n = \mathtt{vol}(\mathcal{R}_n)$ depends on the value of $\boldsymbol{x}_1$. Adopting a Bayesian point of view, we can assume that $\boldsymbol{x}^*$ has a prior distribution μ_0 on $\mathcal{R}_0$, and $g_1(\cdot)$ thus induces some distribution μ for $\boldsymbol{x} = \boldsymbol{x}_1$. Assume that μ is equivalent to the Lebesgue measure.

Note that the dynamical process $(\boldsymbol{z}_n)$ may depend on some other characteristics, *i.e.*, $\boldsymbol{z}_n = (\boldsymbol{x}_n, \boldsymbol{\theta}_n)$, with $\boldsymbol{\theta}_n$ corresponding to characteristics of the objective function $f(\cdot)$ or to parameters of the algorithm. Thus $\boldsymbol{\theta}_1$ is known. The expectations considered below are then with respect to $\boldsymbol{x}_1$ conditional on $\boldsymbol{\theta}_1$. In the cases of root-finding for a nonlinear function and optimisation of a locally symmetric function with a second-order line-search algorithm, the dynamical process depends on the (non-parametric) objective function itself. This can then be written as

$$f(x) = h(x - x^*),$$

with $h(0) = 0$ for root-finding and $h(0) = \min h(\cdot)$ for optimisation. When taking expectations, we then consider $h(\cdot)$ as fixed and perform expectations with respect to x^* only.

We can define the following performance characteristics, assuming the expectations exist:

$$\mathbf{E}\log \mathbf{L}_n = \mathbf{E}_{\boldsymbol{x}}\{\log L_n(\boldsymbol{x})\}, \tag{4.5}$$

$$\log \mathbf{E}\mathbf{L}_n = \log \mathbf{E}_{\boldsymbol{x}}\{L_n(\boldsymbol{x})\}, \tag{4.6}$$

and more generally

$$\log \mathbf{E}\mathbf{L}_n^\gamma = \log \mathbf{E}_{\boldsymbol{x}}\{L_n^\gamma(\boldsymbol{x})\}$$

for $\gamma \geq -1$. We shall also consider the asymptotic versions of these characteristics:

$$W_1 = -\lim_{n\to\infty} \frac{1}{n}\mathbf{E}\log \mathbf{L}_n, \tag{4.7}$$

$$W_2 = -\lim_{n\to\infty} \frac{1}{n}\log \mathbf{E}\mathbf{L}_n, \tag{4.8}$$

and more generally

$$W_\gamma = \frac{1}{1-\gamma}\lim_{n\to\infty} \frac{1}{n}\log \mathbf{E}\mathbf{L}_n^{\gamma-1}, \quad \gamma \geq 0, \gamma \neq 1, \tag{4.9}$$

if the limits exist. We believe that W_γ as a function of γ is a key indicator of the asymptotic behaviour of the algorithm; see Figures 5.1 and 5.14 for examples.

Note that from Jensen's inequality,

$$\log \mathbf{E}\mathbf{L}_n \geq \mathbf{E}\log \mathbf{L}_n, \ \forall n \geq 1,$$

and thus $W_1 \geq W_2$ when W_1 and W_2 exist. More generally, Hölder's inequality implies $W_\gamma \geq W_{\gamma'}$ for $\gamma \leq \gamma'$, when W_γ and $W_{\gamma'}$ exist.

We shall see in Section 4.5.5 that for some line-search algorithms, the characteristic W_γ coincides with the entropy h_γ of the associated dynamical system. Note that $\mathbf{E}\mathbf{L}_n$ is a more natural characteristic of performance for an algorithm than $\mathbf{E}\log \mathbf{L}_n$. Therefore, W_2 seems more appealing than W_1 as a characteristic of asymptotic precision of the location of $\boldsymbol{x}^*$. This opinion will be reinforced in Section 5.7.2 where we shall see that algorithms with totally different asymptotic behaviours have the same value of W_1, whereas the values of W_2 are significantly different.

We shall also consider the performance characteristics related to quantiles of the distribution of $L_n(\boldsymbol{x}_1)$; that is,

$$\bar{L}_n^{1-\alpha} = \sup\{t \ / \ \mathtt{Pr}(L_n(\boldsymbol{x}_1) \geq t) > \alpha\}\,, \tag{4.10}$$

and

$$P_n(\beta) = \mathtt{Pr}(-\frac{1}{n}\log L_n(\boldsymbol{x}_1) < \beta)\,.$$

In the case of second-order line-search algorithms, we shall consider

$$P_n^{GS} = \mathtt{Pr}(L_n(\boldsymbol{x}_1) < L_0\varphi^{n-1})\,, \tag{4.11}$$

and

$$P_n^F = \mathtt{Pr}(L_n(\boldsymbol{x}_1) < \frac{L_0}{F_{n+1}})\,, \tag{4.12}$$

which correspond respectively to the probability of a faster convergence (in terms of $L_n(x_1)$) than the GS and Fibonacci algorithms.

We have the following results.

Theorem 4.2 *Assume that the log-rate ϱ defined by (4.3) exists, and that there exists $c \in (0,1)$ such that for almost all $\boldsymbol{x}_1$, $L_n(\boldsymbol{x}_1) \geq L_1 c^n$; then W_1 exists and equals ϱ.*

Proof. Since for all $n \geq 1$, $L_n(\boldsymbol{x}_1) \leq L_1(\boldsymbol{x}_1) = L_1$ and from the assumption of the theorem,

$$0 \leq -\frac{1}{n}\log\left(\frac{L_n(\boldsymbol{x}_1)}{L_1}\right) \leq -\log c\,.$$

The Lebesgue dominated convergence theorem then implies

$$\begin{aligned} W_1 &= -\lim_{n\to\infty}\frac{1}{n}\mathtt{E}_{\boldsymbol{x}}\{\log L_n(\boldsymbol{x})\} = -\lim_{n\to\infty}\mathtt{E}_{\boldsymbol{x}}\{\frac{1}{n}\log L_n(\boldsymbol{x})\} \\ &= \mathtt{E}_{\boldsymbol{x}}\{-\lim_{n\to\infty}\frac{1}{n}\log L_n(\boldsymbol{x})\} = \varrho\,. \end{aligned}$$

■

Note that $L_n(\boldsymbol{x}_1) \geq L_1 c^n$ for almost all $\boldsymbol{x}_1$ corresponds to the assumption that the rate of convergence is at most exponential for almost all $\boldsymbol{x}_1$.

Using elementary properties of convergence in probability we have

Theorem 4.3 *Assume that the log-rate ϱ defined by (4.3) exists, then $\forall\alpha \in (0,1)$*

$$\lim_{n\to\infty} -\frac{1}{n}\log\bar{L}_n^{1-\alpha} = \varrho$$

and for all $\beta < \varrho$

$$\lim_{n\to\infty} P_n(\beta) = 0\,.$$

4.3 Characteristics of worst-case performances

Considering the worst-case performances with respect to $\boldsymbol{x}_1 \in \mathcal{R}_1$, define the following performance characteristics:

$$\mathtt{ML}_n = \sup_{\boldsymbol{x}_1 \in \mathcal{R}_1} L_n(\boldsymbol{x}_1)\,,$$

and, provided that the limit exists,

$$W_\infty = -\lim_{n\to\infty} \frac{1}{n} \log \mathtt{ML}_n\,. \tag{4.13}$$

Theorem 4.4 *Assume that W_∞ exists. Then for any $\beta < W_\infty$, there exists n_0 such that for all $n \geq n_0$, $P_n(\beta) = 0$.*

Indeed, for any $\delta > 0$, there exists $n_0(\delta)$ such that

$$\forall n \geq n_0(\delta)\,,\ \forall \boldsymbol{x} \in \mathcal{R}_1\,,\ -\frac{1}{n}\log L_n(\boldsymbol{x}) > W_\infty - \delta\,.$$

Taking $\delta = W_\infty - \beta > 0$, we thus obtain $P_n(\beta) = 0$ for all $n \geq n_0(\delta)$.

4.4 Counting characteristic

The uncertainty region $\mathcal{R}_n$ obtained at iteration n depends on $\boldsymbol{x}^*$. However, certain values of $\boldsymbol{x}^*$ lead to the same region $\mathcal{R}_n(\boldsymbol{x}^*)$. (These are exactly the consistency regions.) One can thus count the number of different regions at iteration n that can be obtained when $\boldsymbol{x}^* \in \mathcal{R}_0$. We shall denote this number by $\mathtt{N}_n$. Note that it may depend on the way the algorithm is initialised. We shall also consider the asymptotic version of this characteristic:

$$\varrho_0 = \lim_{n\to\infty} \frac{1}{n} \log \mathtt{N}_n\,, \tag{4.14}$$

provided the limit exists. We show in Section 4.5.5 that for some line-search algorithms ϱ_0 coincides with the topological entropy h_0 of the dynamical system associated with the algorithm. From the definition of W_0, see (4.9), we also have $W_0 = \varrho_0$.

4.5 One-dimensional piecewise linear mappings

One can refer to Section 8.2 (Section 8.2.4 in particular) for classical results in ergodic theory.

4.5.1 Markov-chain representation

Let $T : [0,1) \mapsto [0,1)$ be a piecewise linear mapping; that is, there exist $0 = t_0 < t_1 < \cdots < t_m = 1$ such that $T(\cdot)$ is linear on each interval $[t_{j-1}, t_j)$: $T(x) = a_j x + b_i$ for $x \in [t_{j-1}, t_j)$, $j = 1, \ldots, m$. Examples of such mappings are presented in Figures 3.2, 3.3, 3.8, 5.11. Consider the dynamical system $x_{n+1} = T(x_n)$ and assume that the mapping $T(\cdot)$ is *expanding*, that is, $a_j > 1$ for all $j = 1, \ldots, m$.

Denote $\mathcal{A} = \{t_j\}_{j=0}^m$ and assume that the set of points

$$\mathcal{S}^\infty = \cup_{n=0}^\infty T^n(\mathcal{A})$$

is finite, where $T^n(\mathcal{A}) = \{T^n(x)\,,\ x \in \mathcal{A}\}$. We thus have $\mathcal{S}^\infty = \{\alpha_i\}_{i=0}^M$ for some $M \geq m$, and $0 = \alpha_0 < \alpha_1 < \cdots < \alpha_M = 1$. $\mathcal{S}^\infty$ defines a partition of $[0,1)$ into intervals $\mathcal{I}_i = [\alpha_{i-1}, \alpha_i)$, $i = 1, \ldots, M$. We show now that the study of different performance characteristics of the dynamical system (x_n) can be reduced to the study of certain functionals on finite Markov chains.

Indeed, the dynamical system (x_n) can be given a Markov-chain representation. The intervals $\mathcal{I}_i$, $i = 1, \ldots, M$ are considered as states S_i of the Markov chain. The initial distribution, denoted by $\boldsymbol{\pi}^{(1)}$, is that induced on the states by the prior distribution of x^*. Assume that x_1 has a distribution with constant density on each $\mathcal{I}_i$, then

$$\pi_i^{(1)} = [\boldsymbol{\Pi}^{(1)}]_i = \mathtt{Pr}(x_1 \in \mathcal{I}_i)\,, \quad i = 1, \ldots, M\,.$$

For example, this holds for the line-search algorithms of Section 2.2 when x^* has a uniform prior distribution on $[A_0, B_0]$. For any $n > 1$ the distribution of x_n conditional on $x_n \in \mathcal{I}_i$ is uniform on $\mathcal{I}_i$, and we denote by $\pi_i^{(n)}$ the probability $\mathtt{Pr}(x_n \in \mathcal{I}_i)$. By the definition of $\mathcal{S}^\infty$,

$$T(\mathcal{I}_i) = \cup_{j \in \Omega_i} \mathcal{I}_j\,,$$

where

$$\Omega_i = \{j \;/\; \mathcal{I}_j \subseteq T(\mathcal{I}_i)\} \subseteq \{1, \ldots, M\}\,.$$

We thus define the transition probability from $\mathcal{I}_i$ to $\mathcal{I}_j$ as

$$\pi_{ij} = \begin{cases} \dfrac{|\mathcal{I}_j|}{\sum_{k\in\Omega_i} |\mathcal{I}_k|} & \text{if } j \in \Omega_i \\ 0 & \text{otherwise} \end{cases} \tag{4.15}$$

where $|\mathcal{I}|$ is the length of the interval $\mathcal{I}$. We shall denote by $\boldsymbol{\Pi}$ the matrix with elements π_{ij}.

Ergodicity of $T(\cdot)$ is equivalent to *irreducibility* of the Markov chain; see **H3** in Section 4.5.4. The eigenvector associated with the *simple* eigenvalue 1 for the matrix $\boldsymbol{\Pi}^T$ gives the invariant distribution for the Markov chain, denoted by $\bar{\pi}_i = [\bar{\boldsymbol{\pi}}]_i$, $i = 1, \ldots, M$, with $\sum_{i=1}^{M} \bar{\pi}_i = 1$. The invariant density for x is a mixture of uniform densities in each interval:

$$\phi_T(x) = \sum_{i=1}^{M} \phi_i \mathrm{I}_{\mathcal{I}_i}(x),$$

where $\phi_i = \bar{\pi}_i / |\mathcal{I}_i|$ and $\mathrm{I}_{\mathcal{X}}(\cdot)$ is the indicator function of the set $\mathcal{X}$.

The above measure can also be obtained through the solution of the Frobenius-Perron equation, without requiring a uniform prior distribution for x_1 on each $\mathcal{I}_i$. Indeed, assume that the mapping $T(\cdot)$ is expanding, which will be the case for the dynamical systems associated with line-search algorithms of Section 2.2. Then there exists a T-invariant absolutely continuous measure μ_T. Since in general there exists a j such that $[0, 1) \setminus T([\alpha_{j-1}, \alpha_j)) \neq \emptyset$, its density ϕ_T is discontinuous. The density ϕ_T belongs to $L_1([0, 1], \mu_{\mathcal{L}})$, where $\mu_{\mathcal{L}}$ is the Lebesgue measure. It can therefore be normalised and considered as a probability density. In what follows we assume without loss of generality that $\int_0^1 \phi_T(x) dx = 1$. The invariant density ϕ_T is the eigenfunction of the Frobenius-Perron operator with the eigenvalue 1:

$$\phi_T(x) = \sum_{y_j \in \mathcal{J}(x)} \frac{1}{|T'(y_j)|} \phi_T(y_j), \tag{4.16}$$

where $\mathcal{J}(x) = T^{-1}(x) = \{y_j \;/\; T(y_j) = x\}$, and $T'(\cdot)$ is the derivative of the mapping $T(\cdot)$. From the expanding property of $T(\cdot)$, the invariant measure μ_T is *ergodic*. Ergodicity implies that 1 is a simple eigenvalue of the Frobenius-Perron operator. Thus ϕ_T is the (unique) *ergodic density* associated with the mapping $T(\cdot)$.

Since $T(\cdot)$ is piecewise linear, and in view of the definition of $\mathcal{S}^\infty$, ϕ_T is piecewise constant on the intervals $\mathcal{I}_i$: $\phi_T(x) = \bar{\phi}_i, x \in \mathcal{I}_i, i = 1, \ldots, M$. When M is finite, Equation (4.16) then reduces to a linear equation $\bar{\boldsymbol{\phi}} = \boldsymbol{M}\bar{\boldsymbol{\phi}}$, with $\bar{\boldsymbol{\phi}} = (\bar{\phi}_1, \ldots, \bar{\phi}_M)^T$, and

$$[\boldsymbol{M}]_{ij} = \pi_{ji} \frac{|\mathcal{I}_j|}{|\mathcal{I}_i|},$$

which is the contraction rate from state S_j to state S_i; see (4.15) and (4.17). The equation $\bar{\phi} = M\bar{\phi}$ thus gives

$$\bar{\phi}_i|\mathcal{I}_i| = \sum_{j=1}^{M} \pi_{ji}\bar{\phi}_j |\mathcal{I}_j| ,$$

or equivalently

$$\bar{\pi} = \Pi^T \bar{\pi} ,$$

with $\bar{\pi}_i = \bar{\phi}_i|\mathcal{I}_i|$. This means that $\bar{\pi}$ is the invariant distribution for the Markov chain with matrix Π, given by the normalised eigenvector associated with the eigenvalue 1 for the matrix Π^T.

4.5.2 Finite-sample performance characteristics

In order to be able to base our study of the performances of the algorithm on that of the dynamical system defined by $T(\cdot)$, we need to know the convergence rate obtained at each iteration. This is given by $r_n = 1/|T'(x_n)|$; that is, r_n corresponds to the inverse of the modulus of the slopes of the piecewise linear mapping $T(\cdot)$, a term which appears in the Frobenius-Perron equation.

A reduction rate R_j is thus associated with each state S_j; that is, each interval $\mathcal{I}_j$: R_j is the inverse of the modulus of the slope of the transformation $T(\cdot)$ on the interval $\mathcal{I}_j$,

$$R_j = \frac{|\mathcal{I}_j|}{\sum_{k \in \Omega_j} |\mathcal{I}_k|} . \tag{4.17}$$

From the values of the rates R_j we can easily compute the analytical expressions for $\mathrm{E}\log \mathrm{L}_n$ and $\log \mathrm{E}\mathrm{L}_n$, respectively defined by (4.5) and (4.6), expressed in the following theorem. We shall require the following assumptions.

H1 The partition of $[0, 1)$ defined by $\mathcal{S}^\infty$ is finite.

H2 The prior distribution of x^* is such that the distribution of x_1 is uniform on each interval $\mathcal{I}_i$, $i = 1, \ldots, M$, which induces the initial distribution $\pi^{(1)}$ for the Markov chain.

Theorem 4.5 *Under assumptions* **H1** *and* **H2** *for all* γ *we have*

$$\log \mathrm{E} L_n^\gamma = \gamma \log L_0 + \log p_\gamma^T \Pi_\gamma^{n-2} q_\gamma , \tag{4.18}$$

with

$$[p_\gamma]_i = \pi_i^{(1)} |\mathcal{I}_i|^\gamma , \; [q_\gamma]_i = \frac{R_i^\gamma}{|\mathcal{I}_i|^\gamma} , \; [\Pi_\gamma]_{ij} = (\pi_{ij})^{1+\gamma} , \tag{4.19}$$

and

$$\mathbf{E}\log L_n = \log L_0 + \boldsymbol{\pi}^{(1)^T} \boldsymbol{Q}_{n-2} \boldsymbol{l}\,, \tag{4.20}$$

with

$$\boldsymbol{Q}_{n-2} = \sum_{k=0}^{n-2} \boldsymbol{\Pi}^k\,,\ \boldsymbol{l} = (\log R_1, \ldots, \log R_M)^T\,. \tag{4.21}$$

Proof. First note that $L_n(x_1) = L_0 R_{i_1} R_{i_2} \ldots R_{i_{n-1}}$, where $i_1, \ldots, i_{n-1}$ denotes the sequence of states visited by $x_1, \ldots, x_{n-1}$, which is a function of x_1. The probability of visiting this particular sequence of states is

$$\Pr(x_1 \in \mathcal{I}_{i_1}, x_2 \in \mathcal{I}_{i_2}, \ldots, x_{n-1} \in \mathcal{I}_{i_{n-1}}) = \pi^{(1)}_{i_1} \pi_{i_1 i_2} \ldots \pi_{i_{n-2} i_{n-1}}\,.$$

We thus have

$$\begin{aligned} \log \mathbf{E}\mathrm{L}_n^\gamma &= \log[L_0^\gamma \sum_{i_1,\ldots,i_{n-1}} \pi^{(1)}_{i_1} \pi_{i_1 i_2} \ldots \pi_{i_{n-2} i_{n-1}} \\ &\quad (R_{i_1} R_{i_2} \ldots R_{i_{n-1}})^\gamma]\,. \end{aligned} \tag{4.22}$$

From (4.15) and (4.17), we have

$$R_{i_k} = \frac{\pi_{i_k i_{k+1}} |\mathcal{I}_{i_k}|}{|\mathcal{I}_{i_{k+1}}|}\,, \forall i_{k+1} \in \Omega_{i_k}\,,$$

and substituting back in (4.22) from $k = 1$ to $k = n-2$ we obtain

$$\begin{aligned} \log \mathbf{E}\mathrm{L}_n^\gamma &= \log[L_0^\gamma \sum_{i_1,\ldots,i_{n-1}} \pi^{(1)}_{i_1} |\mathcal{I}_{i_1}|^\gamma (\pi_{i_1 i_2} \ldots \pi_{i_{n-2} i_{n-1}})^{1+\gamma} \\ &\quad \frac{R^\gamma_{i_{n-1}}}{|\mathcal{I}_{i_{n-1}}|^\gamma}]\,, \end{aligned}$$

which can be written as (4.18).

Using similar arguments, $\mathbf{E}\log \mathrm{L}_n$ can be written as

$$\begin{aligned} \mathbf{E}\log \mathrm{L}_n &= \sum_{i_1,\ldots,i_{n-1}} \pi^{(1)}_{i_1} \pi_{i_1 i_2} \ldots \pi_{i_{n-2} i_{n-1}} \log(L_0 R_{i_1} R_{i_2} \ldots R_{i_{n-1}}) \\ &= \log L_0 + \sum_{i_1} \pi^{(1)}_{i_1} \log R_{i_1} \\ &\quad + \sum_{k=2}^{n-1} \sum_{i_1,\ldots,i_k} \pi^{(1)}_{i_1} \pi_{i_1 i_2} \ldots \pi_{i_{k-1} i_k} \log R_{i_k} \\ &= \log L_0 + \boldsymbol{\pi}^{(1)^T} \boldsymbol{l} + \sum_{k=2}^{n-1} \boldsymbol{\pi}^{(1)^T} \boldsymbol{\Pi}^{k-1} \boldsymbol{l}\,, \end{aligned}$$

which gives the result (4.20). ■

4.5.3 Counting cells and zeta function

Simple counting arguments, analogous to the proof of Theorem 4.5, imply that the number of trajectories of length n initialised in state i is

$$\mathtt{N}_n(i) = \boldsymbol{e}_i^T(\boldsymbol{\Pi}_{-1})^n\mathbf{1}$$

where $\boldsymbol{e}_i$ is the i-th coordinate unit vector and

$$[\boldsymbol{\Pi}_{-1}]_{ij} = (\pi_{ij})^0 = \begin{cases} 1 & \text{if } \pi_{ij} > 0 \\ 0 & \text{if } \pi_{ij} = 0 \end{cases}$$

which follows from (4.19) for $\gamma = -1$.

This implies that the characteristic $\mathtt{N}_n$, defined in Section 4.4 and counting the number of possible trajectories of length n, can be computed by the formula

$$\mathtt{N}_n = \mathbf{1}^T(\boldsymbol{\Pi}_{-1})^n\mathbf{1}. \tag{4.23}$$

Another interesting characteristic is the number of periodic trajectories of period n:

$$p_n = \sum_{i=1}^{M} \boldsymbol{e}_i^T(\boldsymbol{\Pi}_{-1})^n\boldsymbol{e}_i = \text{trace}\,(\boldsymbol{\Pi}_{-1})^n. \tag{4.24}$$

This last characteristic is related to the so-called *zeta function*; see Lind and Marcus (1995), p. 192–193,

$$\zeta(t) = \exp\left(\sum_{n=1}^{\infty}\frac{p_n}{n}t^n\right) = \frac{1}{\det(\boldsymbol{I} - t\boldsymbol{\Pi}_{-1})}. \tag{4.25}$$

Note that $\mathtt{N}_n \geq p_n$ for all n.

4.5.4 Asymptotic performance characteristics

To establish asymptotic properties for the dynamical system we shall need some further assumptions about the transition matrix of the associated Markov chain.

H3 The transition matrix $\boldsymbol{\Pi}$ is irreducible; that is,

$$\forall i,j = 1,\ldots,M\,,\ \exists n \text{ such that } [\boldsymbol{\Pi}^n]_{ij} > 0\,.$$

H4 The transition matrix $\boldsymbol{\Pi}$ is strong mixing; that is,

$$[\boldsymbol{\Pi}^n]_{ij} > 0\,,\ \forall i,j = 1,\ldots,M\,,$$

for all $n \geq$ some n_0.

Theorem 4.5 can be used to obtain the value of the characteristic W_γ in (4.9), for $\gamma \geq 0$, $\gamma \neq 1$.

Theorem 4.6 *Assume* **H1** *and* **H2** *are satisfied. Consider the reduced Markov chain consisting of the states reachable from those states S_i such that the initial probabilities $\pi_i^{(1)} > 0$. Let M' be the number of these states, $\mathcal{I}' \subseteq [0,1)$ be the union of the corresponding subintervals $\mathcal{I}_i'$, $i = 1, \ldots, M'$, and $\boldsymbol{\Pi}'$ denote the transition matrix of this reduced chain. Assume that the matrix $\boldsymbol{\Pi}'_{\gamma-1}$ defined by*

$$[\boldsymbol{\Pi}'_{\gamma-1}]_{ij} = ([\boldsymbol{\Pi}']_{ij})^\gamma \tag{4.26}$$

satisfies the strong-mixing property **H4**. *Then for any $\gamma \geq 0$, $\gamma \neq 1$, W_γ exists and is given by*

$$W_\gamma = \frac{1}{1-\gamma} \log \lambda_{\max}(\boldsymbol{\Pi}'_{\gamma-1}), \tag{4.27}$$

where $\lambda_{\max}(\boldsymbol{M})$ denotes the maximal eigenvalue of the matrix $\boldsymbol{M}$.

Proof. From Theorem 4.5, we have

$$\log \mathbf{E} L_n^{\gamma-1} = (\gamma - 1)\log L_0 + \log(\boldsymbol{p}_{\gamma-1}^T {\boldsymbol{\Pi}'_{\gamma-1}}^{n-2} \boldsymbol{q}_{\gamma-1}),$$

with $\boldsymbol{p}_{\gamma-1}, \boldsymbol{q}_{\gamma-1}$ and $\boldsymbol{\Pi}'_{\gamma-1}$ given by (4.19). Since $\boldsymbol{\Pi}'_{\gamma-1}$ satisfies the strong mixing condition **H4**, then a version of the Frobenius-Perron Theorem, see Section 8.2.1, gives

$$\lim_{n\to\infty} \lambda_{\max}^{-(n-2)} \boldsymbol{p}_{\gamma-1}^T {\boldsymbol{\Pi}'_{\gamma-1}}^{n-2} \boldsymbol{q}_{\gamma-1} = (\boldsymbol{p}_{\gamma-1}^T \boldsymbol{u}_{\gamma-1})(\boldsymbol{q}_{\gamma-1}^T \boldsymbol{v}_{\gamma-1}),$$

where $\lambda_{\max} = \lambda_{\max}(\boldsymbol{\Pi}'_{\gamma-1}) > 0$ is the (simple) maximum eigenvalue of $\boldsymbol{\Pi}'_{\gamma-1}$, $\boldsymbol{u}_{\gamma-1}$ is the associated eigenvector and $\boldsymbol{v}_{\gamma-1}$ is the eigenvector of ${\boldsymbol{\Pi}'_{\gamma-1}}^T$ associated with the same eigenvalue, $\lambda_{\max}$. Moreover, $\boldsymbol{u}_{\gamma-1}$ and $\boldsymbol{v}_{\gamma-1}$ have strictly positive elements and $\boldsymbol{u}_{\gamma-1}^T \boldsymbol{v}_{\gamma-1} = 1$. This implies

$$\lim_{n\to\infty} \frac{1}{n} \log\left(\lambda_{\max}^{-(n-2)} \boldsymbol{p}_{\gamma-1}^T {\boldsymbol{\Pi}'_{\gamma-1}}^{n-2} \boldsymbol{q}_{\gamma-1}\right) = 0,$$

so that

$$\lim_{n\to\infty} \frac{1}{n} \log\left(\boldsymbol{p}_{\gamma-1}^T {\boldsymbol{\Pi}'_{\gamma-1}}^{n-2} \boldsymbol{q}_{\gamma-1}\right) = \log \lambda_{\max}(\boldsymbol{\Pi}'_{\gamma-1}),$$

which gives (4.27). ∎

Theorem 4.7 *Assume that* **H1**, **H2** *and* **H3** *are satisfied. Then* W_1 *exists and*

$$W_1 = -\bar{\boldsymbol{\pi}}^T \boldsymbol{l},$$

where $\boldsymbol{l}$ *is defined by (4.21) and* $\bar{\boldsymbol{\pi}}$ *is the invariant distribution for the Markov chain with original transition matrix* $\boldsymbol{\Pi}$.

Proof. From Theorem 4.5,

$$\frac{\mathrm{E}\log \mathrm{L}_n}{n} = \frac{\log L_0}{n} + \frac{\boldsymbol{\pi}^{(1)^T} \boldsymbol{Q}_{n-2} \boldsymbol{l}}{n},$$

with $\boldsymbol{Q}_{n-2}$ and $\boldsymbol{l}$ given by (4.21). Now, **H3** implies that the Markov chain is ergodic, and therefore $\boldsymbol{\pi}^{(1)^T} \boldsymbol{\Pi}^k \boldsymbol{l}$ tends to $\bar{\boldsymbol{\pi}}^T \boldsymbol{l}$ when k tends to infinity, so that

$$W_1 = -\lim_{n\to\infty} \frac{\mathrm{E}\log \mathrm{L}_n}{n} = -\bar{\boldsymbol{\pi}}^T \boldsymbol{l}.$$

■

4.5.5 Relation with entropies

One can refer to Section 8.2.3 for the definition of the entropy $h_\gamma(T,\mu)$ associated with a dynamical system $T(\cdot)$ and a measure μ. We first prove that $W_\gamma = h_\gamma$ for $\gamma \geq 0$, $\gamma \neq 1$.

Theorem 4.8 *Assume that* **H1** *and* **H2** *are satisfied. Define the reduced Markov transition matrix* $\boldsymbol{\Pi}'$ *as in Theorem 4.6, and assume that it satisfies* **H4**. *Then for any* $\gamma \geq 0$, $\gamma \neq 1$,

$$h_\gamma(T,\mu) = \frac{1}{1-\gamma} \log \lambda_{\max}(\boldsymbol{\Pi}'_{\gamma-1}), \tag{4.28}$$

where $\boldsymbol{\Pi}'_{\gamma-1}$ *is defined by (4.26) and* μ *is the initial measure for* x_1.

Proof. According to the Frobenius-Perron Theorem, for any $\gamma \geq 0, \gamma \neq 1$, the eigenvector $\boldsymbol{v}$ of the matrix $\boldsymbol{\Pi}'_{\gamma-1}$ associated with its maximum eigenvalue $\lambda_{\max}(\boldsymbol{\Pi}'_\gamma) > 0$ is unique, up to a constant multiplier, and contains only positive elements. Consider first the case $\gamma > 0$, and denote by μ_γ the distribution for x_1 corresponding to the normalised vector $\boldsymbol{v}_\gamma = (v_1^{1/\gamma}, \ldots, v_{M'}^{1/\gamma})^T$, $\sum_{i=1}^{M'} v_i^{1/\gamma} = 1$. As the initial partition $\mathcal{P}$, take the partition of $\mathcal{I}$ into the intervals $\mathcal{I}_i = [t_{i-1}, t_i)$, $i = 1, \ldots, M'$, corresponding to the states reachable from the states S_j with $\pi_j^{(1)} > 0$.

Denote $\pi'_{ij} = [\boldsymbol{\Pi}']_{ij}$. For every n we have

$$\begin{aligned} H_\gamma(\mathcal{Q}_n(\mathcal{P}), \mu_\gamma) &= H_\gamma(\boldsymbol{v}_\gamma \boldsymbol{\Pi}'_{\gamma-1} \ldots \boldsymbol{\Pi}'_{\gamma-1}) \\ &= \frac{1}{1-\gamma} \log \sum_{i_1,\ldots,i_n} \left(v_{i_1}^{1/\gamma} \pi'_{i_1 i_2} \ldots \pi'_{i_{n-1} i_n} \right)^\gamma \\ &= \frac{1}{1-\gamma} \log \left(\lambda_{\max}^{n-1}(\boldsymbol{\Pi}'_{\gamma-1}) \sum_i v_i \right) \\ &= \frac{n-1}{1-\gamma} \log \lambda_{\max}(\boldsymbol{\Pi}'_{\gamma-1}) + \frac{1}{1-\gamma} \log \sum_i v_i \,. \end{aligned}$$

Therefore

$$h_\gamma(T, \mathcal{P}, \mu_\gamma) = \lim_{n\to\infty} \frac{1}{n} H_\gamma(\mathcal{Q}_n(\mathcal{P}), \mu_\gamma) = \frac{1}{1-\gamma} \log \lambda_{\max}(\boldsymbol{\Pi}'_{\gamma-1}) \,.$$

Since $\mathcal{P}$ is a generating partition, see Section 8.2.3, $h_\gamma(T, \mu_\gamma) = h_\gamma(T, \mathcal{P}, \mu_\gamma)$, where $h_\gamma(T, \mu_\gamma) = \sup_{\mathcal{P}} h_\gamma(T, \mathcal{P}, \mu_\gamma)$. Since there exists k such that for all $i, j = 1, \ldots, M'$, $[(\boldsymbol{\Pi}')^k]_{ij} > 0$, the initial distribution $\boldsymbol{\pi}'^{(1)} = (\pi_1'^{(1)}, \ldots, \pi_{M'}'^{(1)})^T$ of the reduced Markov chain with transition matrix $\boldsymbol{\Pi}'$ can be assumed such that $\pi_i'^{(1)} > 0$ for all $i = 1, \ldots, M'$ (if this is not the case, we simply need to wait until iteration k). Application of Theorem 8.1 thus yields $h_\gamma(T, \mu) = h_\gamma(T, \mu_\gamma)$, with μ the initial measure for x_1.

Consider now the case of the topological entropy ($\gamma = 0$), and define μ_0 as the measure for x_1 corresponding to the invariant distribution for the matrix $\boldsymbol{\Pi}'_{-1}$. The topological entropy counts the number of trajectories of the dynamical system, and is given by

$$\begin{aligned} H_0(\mathcal{Q}_n(\mathcal{P}), \mu_0) &= \log(\text{number of } i_1, \ldots, i_n \ / \\ &\qquad x_{i_1} \in \mathcal{I}_{i_1}, \ldots, x_{i_n} \in \mathcal{I}_{i_n}) \,. \end{aligned}$$

Considering the same reduced mapping $T_r(\cdot)$ as above, we get

$$H_0(\mathcal{Q}_n(\mathcal{P}), \mu_0) = \log \left(\sum_{i_1,\ldots,i_n} [\boldsymbol{\Pi}'_{-1}]_{i_1 i_2} \ldots [\boldsymbol{\Pi}'_{-1}]_{i_{n-1} i_n} \right) \,.$$

We then obtain

$$h_0(T, \mu_0) = \lim_{n\to\infty} \frac{1}{n} \log(\mathbf{1}^T (\boldsymbol{\Pi}'_{-1})^{n-1} \mathbf{1}) \,,$$

which gives $h_0(T, \mu) = h_0(T, \mu_0) = \log \lambda_{\max}(\boldsymbol{\Pi}'_{-1})$. ∎

From Theorems 4.6 and 4.8, we thus have $W_\gamma = h_\gamma$ for $\gamma \geq 0$, $\gamma \neq 1$.

The analogue of (4.28) for $\gamma = 1$ is well known; see Petersen (1983), p. 246:

$$h_1(T, \mu_T) = -\sum_{i,j} \bar{\pi}_i \pi_{ij} \log[\boldsymbol{\Pi}]_{ij} , \tag{4.29}$$

where μ_T is the measure for x_1 corresponding to the vector $\bar{\boldsymbol{\pi}} = (\bar{\pi}_1, \ldots, \bar{\pi}_N)^T$ of stationary probabilities of the transition matrix $\boldsymbol{\Pi}$. In information theory, the analogue of the metric entropy $h_1(T, \mu_T)$ is called the entropy rate of a stochastic process, and (4.29) gives the entropy rate of a stationary Markov chain. We then get the following.

Theorem 4.9 *Assume that* **H1**, **H2** *and* **H3** *are satisfied; then* $W_1 = h_1(T, \mu_T)$.

Proof. $h_1(T, \mu_T)$ can be written as

$$h_1(T, \mu_T) = -\sum_{i,j} \bar{\pi}_i \pi_{ij} \log \pi_{ij} ,$$

and thus from (4.15, 4.17),

$$\begin{aligned} h_1(T, \mu_T) &= -\sum_{i,j} \bar{\pi}_i \pi_{ij} \log \left(\frac{|\mathcal{I}_j|}{|\mathcal{I}_i|} R_i \right) , \\ &= -\sum_{i,j} \bar{\pi}_i \pi_{ij} \log R_i - \sum_{i,j} \bar{\pi}_i \pi_{ij} \log(|\mathcal{I}_j|) \\ &\quad + \sum_{i,j} \bar{\pi}_i \pi_{ij} \log(|\mathcal{I}_i|) \\ &= -\sum_{i} \bar{\pi}_i \log R_i , \end{aligned}$$

which equals W_1 from Theorem 4.7. ∎

Another proof of Theorem 4.9 follows from the Shannon-McMillan-Breiman Theorem; see Petersen (1983) p. 261: for almost all x^* in $[0, 1)$, $-(1/n) \log L_n(x^*)$ tends to $h_1(T, \mu_T)$ as n tends to infinity. Theorem 4.2 then implies that $h_1(T, \mu_T) = \varrho = W_1$.

CHAPTER 5

Line-search algorithms

5.1 First-order line search

5.1.1 Behaviour for linear functions

Markov-chain representation and subshifts of finite type

Consider the family of algorithms defined by (2.3). When the function $f(\cdot)$ is linear, one can associate with the algorithm the dynamical system (3.2). Consider then the sets

$$\mathcal{A} = \{0, e, 1/2, 1-e, 1\}\,,\ \mathcal{S}^\infty = \cup_{n=0}^{\infty} T^n(\mathcal{A})\,,$$

where $T^n(\mathcal{A}) = \{T^n(z)\,,\ z \in \mathcal{A}\}$. $\mathcal{S}^\infty$ defines a partition of $[0,1]$ into closed intervals $\mathcal{I}_i$, $i = 1, \ldots, M \leq \infty$, which can be considered as states of the dynamical system. Note that the partition can be made infinite by choosing e as a non-algebraic number; see also Example 5.5 in Section 5.7. We shall consider different values e that give a finite partition $\mathcal{S}^\infty$. These values are obtained by considering the successive iterates of e by the transformation $T(\cdot)$; see Figure 3.3. For example,

(i) take $T[(1/2)^-] = e$, it gives $e = 1 - \sqrt{2}/2 \simeq 0.29289$;

(ii) take $T[(1/2)^-] = e' < e$ and $T(e') = 1/2$, it gives $e = 1 - \varphi \simeq 0.381966$;

(iii) take $T[(1/2)^-] = e' < e$ and $T(e') = 1 - e'$, it gives $e = 1/3$;

(iv) take $T[(1/2)^-] = e' < e$ and $T(e') = e$, which gives e as a solution of $2e^3 - 2e^2 - 2e + 1 = 0$, that is, $e \simeq 0.40303$.

From the results presented in Section 4.5, this dynamical process has an invariant measure μ_T absolutely continuous with respect to the Lebesgue measure.

Example 5.1 *Take $e = 1 - \sqrt{2}/2$ in (3.2). This gives the partition*

$$\mathcal{I}_1 = [0, e)\,,\ \mathcal{I}_2 = [e, 1/2)\,,\ \mathcal{I}_3 = [1/2, 1-e)\,,\ \mathcal{I}_4 = [1-e, 1)\,,$$

and the mapping $T(\cdot)$ satisfies:

$$\begin{aligned}
\mathcal{I}_1 &\mapsto \mathcal{I}_1 \cup \mathcal{I}_2 \cup \mathcal{I}_3 \cup \mathcal{I}_4 \\
\mathcal{I}_2 &\mapsto \mathcal{I}_1 \\
\mathcal{I}_3 &\mapsto \mathcal{I}_4 \\
\mathcal{I}_4 &\mapsto \mathcal{I}_1 \cup \mathcal{I}_2 \cup \mathcal{I}_3 \cup \mathcal{I}_4
\end{aligned}$$

The transition matrix $\boldsymbol{\Pi}$ is given by

$$\boldsymbol{\Pi} = \begin{pmatrix} e & 1/2-e & 1/2-e & e \\ 1 & 0 & 0 & 0 \\ 0 & 0 & 0 & 1 \\ e & 1/2-e & 1/2-e & e \end{pmatrix},$$

and the invariant distribution is

$$\bar{\boldsymbol{\pi}} = [1/(2\sqrt{2})] \times (1, 1-2e, 1-2e, 1)^T .$$

Example 5.2 *Take $e = 1 - \varphi$. This gives the partition*

$$\begin{aligned}
&\mathcal{I}_1 = [0, (1-\varphi)/2), && \mathcal{I}_2 = [(1-\varphi)/2, 1-\varphi), \\
&\mathcal{I}_3 = [1-\varphi, 1/2), && \mathcal{I}_4 = [1/2, \varphi), \\
&\mathcal{I}_5 = [\varphi, (1+\varphi)/2), && \mathcal{I}_6 = [(1+\varphi)/2, 1);
\end{aligned}$$

the transition matrix

$$\boldsymbol{\Pi} = \begin{pmatrix} 1-\varphi & 1-\varphi & 2\varphi-1 & 0 & 0 & 0 \\ 0 & 0 & 0 & 2\varphi-1 & 1-\varphi & 1-\varphi \\ 1 & 0 & 0 & 0 & 0 & 0 \\ 0 & 0 & 0 & 0 & 0 & 1 \\ 1-\varphi & 1-\varphi & 2\varphi-1 & 0 & 0 & 0 \\ 0 & 0 & 0 & 2\varphi-1 & 1-\varphi & 1-\varphi \end{pmatrix};$$

and the invariant distribution

$$\bar{\boldsymbol{\pi}} = (1/4) \times (1, \varphi, 1-\varphi, 1-\varphi, \varphi, 1)^T .$$

Example 5.3 *Take $e = 1/3$. This gives the partition*

$$\begin{aligned}
&\mathcal{I}_1 = [0, 1/4), && \mathcal{I}_2 = [1/4, 1/3), && \mathcal{I}_3 = [1/3, 1/2), \\
&\mathcal{I}_4 = [1/2, 2/3), && \mathcal{I}_5 = [2/3, 3/4), && \mathcal{I}_6 = [3/4, 1);
\end{aligned}$$

the transition matrix

$$\Pi = \begin{pmatrix} 1/3 & 1/9 & 2/9 & 2/9 & 1/9 & 0 \\ 0 & 0 & 0 & 0 & 0 & 1 \\ 1 & 0 & 0 & 0 & 0 & 0 \\ 0 & 0 & 0 & 0 & 0 & 1 \\ 1 & 0 & 0 & 0 & 0 & 0 \\ 0 & 1/9 & 2/9 & 2/9 & 1/9 & 1/3 \end{pmatrix};$$

and the invariant distribution

$$\bar{\pi} = (1/15) \times (9/2, 1, 2, 2, 1, 9/2)^T .$$

Performance characteristics

Worst-case performances can be studied on a case-by-case basis. For instance, when $e = 1 - \sqrt{2}/2$ in (3.2), the worst performances are for the cycle $\mathcal{I}_1 \to \mathcal{I}_2 \to \mathcal{I}_1 \to \mathcal{I}_2 \cdots$, which gives

$$\begin{aligned} W_\infty &= -\frac{1}{2}\log[e(1-e)] = -\frac{1}{2}\log(\sqrt{2}-1) + \frac{1}{2}\log 2 \\ &\simeq 0.78726 , \end{aligned}$$

with W_∞ defined by (4.13). This value must be compared to the value $W_\infty = \log 2 \simeq 0.6930$ obtained for the bifurcation algorithm. One can also easily check that starting at $n = 4$ the algorithm defined by (2.3) with $e = 1 - \sqrt{2}/2$ has a better performance than the bifurcation algorithm, whatever the value of x^* (when $f(\cdot)$ is linear).

When $e_1/3$, the worst performances are obtained for the same cycle as above, which gives

$$W_\infty = -\frac{1}{2}\log[e(1-e)] = -\frac{1}{2}\log\frac{2}{9} \simeq 0.75204 ,$$

again a better value than the bifurcation algorithm. One can show that in that case, the algorithm (2.3) has a better performance than the bifurcation algorithm (for linear functions) starting at $n = 6$, whatever the value of x^*.

Average performances can be calculated as shown in Section 4.5. The invariant measure of the dynamical system (3.2) is ergodic. The log-rate ϱ defined by (4.3) thus exists, since the rate r_n as a function of x_n is piecewise constant on $[0, 1)$. One has $r(x) = 1/|T'(x)|$. From Theorem 4.1, ϱ is also the Lyapunov exponent Λ of the dynamical process. From Theorem 4.2, we also have $\varrho = W_1$. The values of ϱ for $e = 1 - \sqrt{2}/2$, $1/3$ and $1 - \varphi$ are indicated in Table 5.1. The values of $\lambda_{\max}(\boldsymbol{\Pi}_{-1})$ and $W_0 = \varrho_0$ (resp. of $\lambda_{\max}(\boldsymbol{\Pi}_1)$ and W_2) are indicated in Table 5.2 (resp. in Table 5.3).

The zeta function, see (4.25), for the three values of e considered is given by:

$$\zeta(t) = \begin{cases} \frac{1}{1-2t-2t^2} & \text{for } e = 1-\sqrt{2}/2 \\ \frac{1}{1-2t-2t^2+2t^3+t^4} & \text{for } e = 1-\varphi \\ \frac{1}{1-2t-3t^2+4t^3} & \text{for } e = 1/3 \end{cases}$$

A Taylor series development of $\log\zeta(t)$ gives the number p_n of periodic trajectories of period n, see (4.24), as given in Table 5.4.

e	$\varrho = W_1$
$1-\frac{\sqrt{2}}{2}$	$-\frac{1}{\sqrt{2}}\left[\log(1-\frac{\sqrt{2}}{2}) + (\sqrt{2}-1)\log\frac{\sqrt{2}}{2}\right] \simeq 0.969799$
$1/3$	$\log 3 - \frac{4}{15}\log 2 \simeq 0.913773$
$1-\varphi$	$\frac{1+\varphi}{2}\log(2+\varphi) + \frac{1-\varphi}{2}\log(1+\varphi) \simeq 0.870520$

Table 5.1 *Values of $\varrho = W_1$ for different choices of e*

e	$\lambda_{\max}(\boldsymbol{\Pi}_{-1})$	$\varrho_0 = W_0$
$1-\frac{\sqrt{2}}{2}$	$1+\sqrt{3} \simeq 2.73205$	1.00505
$1/3$	$1/2+\sqrt{17}/2 \simeq 2.561553$	0.940614
$1-\varphi$	$1+\sqrt{2} \simeq 2.414214$	0.8813736

Table 5.2 *Values of $\varrho_0 = W_0$ for different choices of e*

Figure 5.1 gives the evolution of W_γ as a function of γ for the three values of e that have been considered. Note that W_γ is a decreasing function of γ, see Section 4.2, and that the best performances are when $e = 1-\sqrt{2}/2$, whatever the value of γ.

The bifurcation algorithm of Sections 2.2.1 and 3.2.1 is such that $W_\gamma = \varrho = \log 2$ for any $\gamma \geq 0$. This means in particular that the algorithm (2.3) with $e = 1-\sqrt{2}/2$ has better performances than the bifurcation algorithm in the sense of W_γ for any γ for linear functions. The next section gives a bound on the best achievable value of ϱ.

e	$\lambda_{\max}(\boldsymbol{\Pi}_1)$	W_2
$1-\frac{\sqrt{2}}{2}$	$\frac{3}{2}-\sqrt{2}+\frac{1}{2}\sqrt{23-16\sqrt{2}}$	0.939088
$1/3$	$\frac{1}{18}(1+\sqrt{41})$	0.888470
$1-\varphi$	$\frac{7}{2}-\frac{3}{2}\sqrt{5}+\frac{1}{2}\sqrt{130-58\sqrt{5}}$	0.859408

Table 5.3 *Values of W_2 for different choices of e*

e	p_1	p_2	p_3	p_4	p_5	p_6	p_7	p_8	p_9	p_{10}
$1-\frac{\sqrt{2}}{2}$	2	8	20	56	152	416	1136	3104	8480	23168
$1-\varphi$	2	8	14	36	82	200	478	1156	2786	6728
$1/3$	2	10	14	50	102	298	702	1890	4694	12250

Table 5.4 *Number of periodic trajectories of period n for different choices of e, $n = 1, \ldots, 10$*

An upper bound for the log-rate

Assume that the function $f(\cdot)$ is linear on $[A_0, B_0)$, with $f(A_0) < 0 < f(B_0)$. Let $f_n(\cdot)$ denote the renormalised function on $[0,1)$ at iteration n, and assume that $|f_n(0)| < |f_n(1)|$. This gives $x_n < 1/2$. We thus obtain an optimistic interval $[0, 1/2)$ for the location of x_n, based on the assumption of linearity. We then renormalise this interval to $[0,1)$, and $f_{n+1}(0)$ is already known. Next, we evaluate $f_{n+1}(\cdot)$ at $e_{n+1} = \alpha$ in $[0,1)$. Three cases must be distinguished.

(i) $f_{n+1}(e_{n+1}) < 0$, which implies $x_{n+1} \in [e_{n+1}, 1)$,

(ii) $0 < f_{n+1}(e_{n+1})$ and $|f_{n+1}(0)| \geq f_{n+1}(e_{n+1})$, which implies $x_{n+1} \in [e_{n+1}/2, e_{n+1})$,

(iii) $0 \leq f_{n+1}(e_{n+1})$ and $|f_{n+1}(0)| < f_{n+1}(e_{n+1})$, which implies $x_{n+1} \in [0, e_{n+1}/2)$.

In each case, the function is already evaluated at one of the end points of the new interval. Similarly, if at iteration n, $|f_n(0)| > |f_n(1)|$, then $x_n \in [1/2, 1)$ which defines the optimistic interval, with $f_{n+1}(1)$ already known. We then evaluate $f_{n+1}(\cdot)$ at $\beta \in [0,1)$. Taking $\alpha = 2/3$ and $\beta = 1/3$ obviously gives $r_n = 1/3$, $\forall n$, and thus $\varrho = \log 3$.

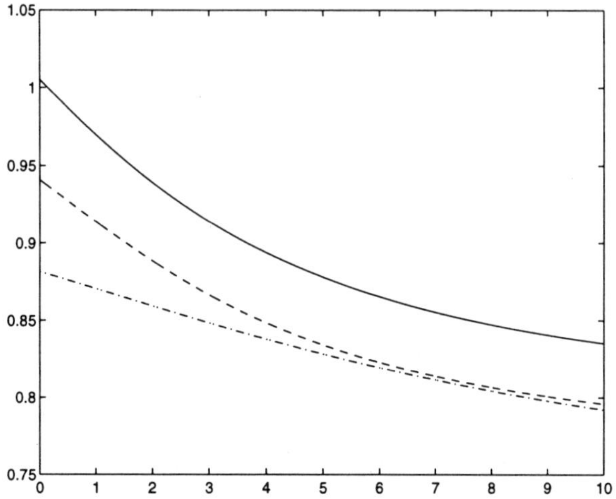

Figure 5.1 *Evolution of W_γ as a function of γ for the algorithm (2.3) when $e = 1 - \sqrt{2}/2$ (full line), $e = 1/3$ (dashed line) and $e = 1 - \varphi$ (dash-dotted line)*

One may then wonder if a better log-rate can be achieved for x^* lying in a set of positive Lebesgue measure. The answer is no, as shown below.

Theorem 5.1 *For any sequence (e_k), $0 < e_k < 1$, $k = 0, 1, \ldots$*

$$\mu_{\mathcal{L}}\{x_1 \in [0,1) \;/\; \liminf_{n\to\infty}[L_n(x_1)]^{\frac{1}{n}} < \frac{1}{3}\} = 0\,,$$

where $\mu_{\mathcal{L}}$ is the Lebesgue measure and $L_n(x_1) = L_0 r_0 r_1 r_2 \ldots r_{n-1}$, with $L_0 = B_0 - A_0$.

Proof. Any first-order line-search algorithm using the optimistic rule is determined by a sequence (e_n). After n iterations, the interval $\mathcal{K}_n$ for x_1 for which the reduction coefficients coincide with $(r_k)_{k=0}^{n-1}$ has the length $L_n(x_1)$. There are 3^n such intervals at iteration n, due to the three cases (i), (ii) and (iii) above.

Let us fix δ, $0 < \delta < 1/3$ and consider the sets

$$\mathcal{X}_{n,\delta} = \{x_1 \in [0,1) \;/\; [L_n(x_1)]^{\frac{1}{n}} < \frac{1}{3} - \delta\}\,.$$

The total length of the union of the sub-intervals $\mathcal{K}_n \cap \mathcal{X}_{n,\delta}$ cannot exceed the number of disjoint intervals $\mathcal{K}_n$ times the value of L_n

and is therefore bounded above by $3^n(1/3 - \delta)^n$. Thus, applying Borel-Cantelli arguments

$$\mu_{\mathcal{L}}\{x_1 \in [0,1) \; / \; \liminf_{n\to\infty}[L_n(x_1)]^{\frac{1}{n}} < \frac{1}{3} - \delta\} \; = $$
$$\mu_{\mathcal{L}}\{x_1 \in [0,1) \; / \; [L_n(x_1)]^{\frac{1}{n}} < \frac{1}{3} - \delta \text{ infinitely often }\} \; = \; 0\,,$$

since for any $0 < \delta < 1/3$

$$\sum_{n=1}^{\infty} \mu_{\mathcal{L}}\{x_1 \in [0,1) \; / \; [L_n(x_1)]^{\frac{1}{n}} < \frac{1}{3} - \delta\} \leq \sum_{n=1}^{\infty} 3^n(\frac{1}{3} - \delta)^n < \infty\,.$$

This completes the proof. ∎

5.1.2 Behaviour for locally linear functions

We show in this section that the asymptotic behaviour of the dynamical system (3.2) characterized in the previous section for a linear function $f(\cdot)$ is the same when $f(\cdot)$ is only locally linear at x^*. Similar results will be obtained in Section 5.3.2 for the GS algorithm: its asymptotic behaviour is the same for a function symmetric about x^* or only locally symmetric. Proofs will be more developed there, and one can thus refer to Section 5.3.2 for more details.

Let (x'_n) denote the sequence of renormalised values of x^* for the actual process driven by $f(\cdot)$ locally linear at x^*. It satisfies:

$$x'_{n+1} = \begin{cases} \frac{x'_n}{e} & \text{if } |f_n(0)| \leq |f_n(1)| \text{ and } f_n(e) \geq 0 \\ \frac{x'_n - e}{1-e} & \text{if } |f_n(0)| \leq |f_n(1)| \text{ and } f_n(e) < 0 \\ \frac{x'_n}{1-e} & \text{if } |f_n(0)| > |f_n(1)| \text{ and } f_n(1-e) \geq 0 \\ \frac{x'_n-(1-e)}{e} & \text{if } |f_n(0)| > |f_n(1)| \text{ and } f_n(1-e) < 0 \end{cases} \tag{5.1}$$

The function $f(\cdot)$ is assumed to satisfy $f(x) = C_1(x - x^*) + \mathcal{O}(|x - x^*|^\beta)$, with $C_1 > 0$ and $\beta > 1$; that is,

$$C_1(x-x^*) - C_2|x-x^*|^\beta \leq f(x) \leq C_1(x-x^*) + C_2|x-x^*|^\beta\,, \tag{5.2}$$

for some $C_1 > 0$, $C_2 \geq 0$, $\beta > 1$.

Since the size of the uncertainty interval $[A_n, B_n)$ is reduced at least by a factor $1 - e$ at each iteration ($e \leq 1/2$), we can choose the renormalised function $f_n(\cdot)$ such that it satisfies

$$C_1(x - x_n) - C_2|x - x_n|^\beta(1-e)^{n(\beta-1)} \leq f_n(x) \leq$$

$$C_1(x - x_n) + C_2|x - x_n|^\beta (1-e)^{n(\beta-1)},$$

so that

$$C_1(x - x_n) - C_2'(1-e)^{n(\beta-1)} \le f_n(x) \le \\ C_1(x - x_n) + C_2'(1-e)^{n(\beta-1)}, \tag{5.3}$$

with $C_2' = C_2 L_1/L_0$.

We shall say that a *mistake* has been made at iteration n of the process (x_k') if $x_n = x_n'$ and $x_{n+1} \neq x_{n+1}'$. We can show that, as n increases, the size of the interval where a mistake can occur decreases exponentially. Indeed, a mistake can occur at iteration n if $|f_n(0)| \le |f_n(1)|$ and $x_n > 1/2$, or $|f_n(0)| > |f_n(1)|$ and $x_n \le 1/2$. Now, from (5.3), $|f_n(0)| \le |f_n(1)|$ is possible only if $C_1 x_n + C_2'(1-e)^{n(\beta-1)} \le C_1(1-x_n) + C_2'(1-e)^{n(\beta-1)}$; that is, $x_n \le 1/2 + (C_2'/C_1)(1-e)^{n(\beta-1)}$. Similarly, $|f_n(0)| > |f_n(1)|$ is possible only if $x_n > 1/2 - (C_2'/C_1)(1-e)^{n(\beta-1)}$, which gives the following condition on x_n to make mistakes possible:

$$x_n \in \left(\frac{1}{2} - \frac{C_2'}{C_1}(1-e)^{n(\beta-1)}, \frac{1}{2} + \frac{C_2'}{C_1}(1-e)^{n(\beta-1)}\right].$$

For some particular values of e, the line-search algorithm possesses a *self-correcting property*; that is, if $x_n = x_n'$ and $x_{n+1} \neq x_{n+1}'$, then after a finite number of iterations, again $x_{n+k} = x_{n+k}'$. Consider for instance the case $e = 1 - \sqrt{2}/2$. The mapping $T(\cdot)$ is presented in Figure 3.3. Take $\epsilon < 1/2 - e$ and $x_n = x_n' = 1/2 - \epsilon$ (the case $x_n = x_n' = 1/2 + \epsilon$ could be treated analogously). Assume that a mistake occurs, which gives

$$x_{n+1} = e - \tfrac{\epsilon}{1-e}, \qquad x_{n+1}' = 1 - e - \tfrac{\epsilon}{1-e},$$
$$x_{n+2} = 1 - \tfrac{\epsilon}{e(1-e)}, \quad x_{n+2}' = 1 - \tfrac{\epsilon}{(1-e)^2},$$

and

$$x_{n+j} = 1 - \frac{\epsilon}{e^{j-1}(1-e)}, \; x_{n+j}' = 1 - \frac{\epsilon}{(1-e)^2 e^{j-2}}$$

until $x_{n+j} < 1 - e$. Then, if $1 - e < x_{n+j}'$,

$$x_{n+j+1} = x_{n+j+1}' = 1 - \frac{\epsilon}{(1-e)^2 e^{j-1}};$$

otherwise

$$x_{n+j+1} = 1 - \frac{\epsilon}{(1-e)^2 e^{j-1}}, \; x_{n+j+1}' = 1 - \frac{\epsilon}{(1-e)^3 e^{j-2}},$$

and the process repeats until for some k: $x_{n+k} < 1 - e < x'_{n+k}$, so that $x_{n+k+1} = x'_{n+k+1}$. One can then prove the following theorem.

Theorem 5.2 *Let $f(\cdot)$ satisfy the condition (5.2) and e be such that the algorithm possesses the self-correcting property above. Then the ergodic behaviours of the dynamical systems (x_n) and (x'_n), defined respectively by (3.2) and (5.1), coincide in the sense that for almost all $x^* \in [A_0, B_0)$ and any Borel set $\mathcal{A} \subset [A_0, B_0)$*

$$\lim_{n\to\infty} \frac{1}{n} \sum_{k=1}^{n} \mathrm{I}_{\mathcal{A}}(x'_k) = \mu_T(\mathcal{A}), \qquad (5.4)$$

where μ_T is the invariant measure for the process (x_n).

Moreover, if the characteristics W_γ of average performances defined by (4.9) exist for the linear approximation of $f(\cdot)$ at x^*, then those of the dynamical process associated with $f(\cdot)$ exist too, and they coincide.

Proof. The proof of (5.4) follows the same lines as the proof of Theorem 5.3 which concerns the GS algorithm. The key points are (i) mistakes are corrected in a finite number of iterations, and (ii) the lengths of intervals for x_n where mistakes can occur decrease exponentially as n increases. The dynamical system (x'_n) defined by (5.1) corresponds to the function $f(\cdot)$, whereas (x_n) defined by (3.2) corresponds to the linear approximation $g(\cdot)$ of $f(\cdot)$. The self-correcting property implies that for any $f(\cdot)$ satisfying (5.2), any x_1 and any n

$$0 < c_1 < \frac{L_n(x_1)}{L'_n(x_1)} < c_2 < \infty ,$$

for some constants c_1, c_2. Then,

$$\log c_1 < \mathrm{E}_x\{\log L_n(x)\} - \mathrm{E}_x\{\log L'_n(x)\} < \log c_2 ,$$

and thus W_1 is the same for $f(\cdot)$ and $g(\cdot)$. Also,

$$c_1^{\gamma-1}\mathrm{E}_x\{[L'_n(x)]^{\gamma-1}\} < \mathrm{E}_x\{[L_n(x)]^{\gamma-1}\} < c_2^{\gamma-1}\mathrm{E}_x\{[L'_n(x)]^{\gamma-1}\}$$

for any $\gamma \geq 0$, $\gamma \neq 1$, and thus the value of W_γ is the same for $f(\cdot)$ and $g(\cdot)$. ■

5.2 Continued fraction expansion and the Gauss map

The ergodic behaviour of the continued fraction approximation derives from those of the Gauss map (ϵ_n). We shall investigate the behaviour of (ϵ_n, s_n, x_n) for the pessimistic and optimistic processes defined in Section 3.2.2, noting that ϵ_n has the invariant density $1/[(1+x)\log 2]$; see (3.3).

We start with the dynamical system describing the evolution of (ϵ_n, s_n), which is common to both cases:

$$\begin{cases} \epsilon_{n+1} &= \frac{1}{\epsilon_n} - \left\lfloor \frac{1}{\epsilon_n} \right\rfloor \\ s_{n+1} &= \frac{1}{s_n + \left\lfloor \frac{1}{\epsilon_n} \right\rfloor} \end{cases}$$

with $\epsilon_1 = x^*$, $s_1 = 0$; see (3.6). The Jacobian of the transformation

$$(\epsilon_n, s_n) \mapsto (\epsilon_{n+1}, s_{n+1}) = T(\epsilon_n, s_n)$$

evaluated at (ϵ, s) is

$$\boldsymbol{J}_T(\epsilon, s) = \begin{pmatrix} -1/\epsilon^2 & 0 \\ 0 & -1/(s + \lfloor 1/\epsilon \rfloor)^2 \end{pmatrix}.$$

The Frobenius-Perron equation for the invariant density $\phi(\epsilon, s)$ is

$$\phi(\epsilon, s) = \sum_{(\epsilon', s') \in T^{-1}(\epsilon, s)} |\boldsymbol{J}_T(\epsilon', s')|^{-1} \phi(\epsilon', s').$$

A useful consequence of the process being two dimensional is that the mapping $T(\cdot)$ is one-to-one, and there is, in fact, no need for the summation above. After some computations we obtain

$$\phi(\epsilon, s) = \frac{1}{s^2(\epsilon + m)^2} \phi[1/(\epsilon + m), (1 - ms)/s],$$

where $m = \lfloor 1/\epsilon \rfloor$. The solution

$$\phi(\epsilon, s) = \frac{1}{(1 + \epsilon s)^2 \log 2}, \quad 0 \le \epsilon, s \le 1, \tag{5.5}$$

which can easily be verified, appears in Dajani and Kraaikamp (1994), together with the proof that for almost all x^* the density (5.5) is indeed the asymptotic density of the two-dimensional sequence (ϵ_n, s_n) with $\epsilon_1 = x^*$ and $s_1 = 0$. Note that the density (3.3) of the Gauss map is easily obtained from (5.5) by integrating $\phi(\epsilon, s)$ with respect to s.

For each of the two cases, pessimistic and optimistic, the properties of (x_n) are inherited from those of (ϵ_n, s_n).

In the pessimistic case, where $x_n = \epsilon_n s_n/(1 + \epsilon_n s_n)$, the asymptotic density for x_n is

$$\phi(x) = \frac{1}{\log 2} \log\left(\frac{1-x}{x}\right), \quad x \in (0, 1/2].$$

This result is similar to Knuth's result, see Knuth (1984), on the asymptotic distribution ($n \to \infty$) of $\theta_n(x) = q_n^2|x - p_n/q_n|$ where p_n/q_n is the n-th continued fraction convergent of a random x uniformly distributed in $[0,1]$ and q_n is the corresponding denominator; see Section 2.2.2.

For the optimistic case, where $x_n = \epsilon_n(1+s_n)/(1+\epsilon_n s_n)$, the asymptotic density for x_n is uniform: $\phi(x) = 1$ for $x \in [0,1]$. This is intuitively clear from observing the propagation of a uniform density on the optimistic interval, which is the consistent set.

An alternative approach is to write down the dynamical systems for (ϵ_n, x_n) directly. In the pessimistic case it is

$$(\epsilon_n, x_n) \mapsto (\epsilon_{n+1}, x_{n+1}) = T(\epsilon_n, x_n),$$

with

$$\begin{cases} \epsilon_{n+1} &= \frac{1}{\epsilon_n} - \left\lfloor \frac{1}{\epsilon_n} \right\rfloor \\ x_{n+1} &= (1-x_n)(1-\epsilon_n \left\lfloor \frac{1}{\epsilon_n} \right\rfloor). \end{cases}$$

In the optimistic case it is

$$(\epsilon_n, x_n) \mapsto (\epsilon_{n+1}, x_{n+1}) = \tilde{T}(\epsilon_n, x_n),$$

with

$$\begin{cases} \epsilon_{n+1} &= \frac{1}{\epsilon_n} - \left\lfloor \frac{1}{\epsilon_n} \right\rfloor \\ x_{n+1} &= \frac{(\epsilon_n \left\lfloor \frac{1}{\epsilon_n} \right\rfloor - 1)(\epsilon_n \left\lfloor \frac{1}{\epsilon_n} \right\rfloor x_n - \epsilon_n \left\lfloor \frac{1}{\epsilon_n} \right\rfloor + \epsilon_n x_n - x_n)}{\epsilon_n(1-\epsilon_n)}. \end{cases}$$

The Jacobians are respectively

$$\boldsymbol{J}_T = \begin{pmatrix} -1/\epsilon_n^2 & 0 \\ (x_n - 1)\left\lfloor \frac{1}{\epsilon_n} \right\rfloor & \epsilon_n \left\lfloor \frac{1}{\epsilon_n} \right\rfloor - 1 \end{pmatrix},$$

and

$$\boldsymbol{J}_{\tilde{T}} =$$

$$\begin{pmatrix} -1/\epsilon_n^2 & 0 \\ \frac{\epsilon_n^2 \left\lfloor \frac{1}{\epsilon_n} \right\rfloor (x_n - 1)(\left\lfloor \frac{1}{\epsilon_n} \right\rfloor - 1) - x_n(1-\epsilon_n)^2}{\epsilon_n^2(1-\epsilon_n)^2} & \frac{(\epsilon_n \left\lfloor \frac{1}{\epsilon_n} \right\rfloor - 1)(\epsilon_n \left\lfloor \frac{1}{\epsilon_n} \right\rfloor + \epsilon_n - 1)}{\epsilon_n(1-\epsilon_n)} \end{pmatrix}.$$

From this, the joint invariant densities for (ϵ_n, x_n) can be found by solving the Frobenius-Perron equations. We obtain respectively

$$\phi(\epsilon, x) = \frac{1}{\epsilon \log 2}, \quad \text{for } 0 \le \epsilon \le 1, \ 0 \le x \le \frac{\epsilon}{1+\epsilon},$$

and

$$\tilde{\phi}(\epsilon, x) = \frac{1}{\epsilon(1-\epsilon)\log 2}, \quad \text{for } 0 \le \epsilon \le 1\,,\ \epsilon \le x \le \frac{2\epsilon}{1+\epsilon}\,.$$

Figure 5.2 (respectively 5.3) presents a plot of a typical sequence of iterates (ϵ_n, x_n) in the pessimistic (respectively optimistic) case.

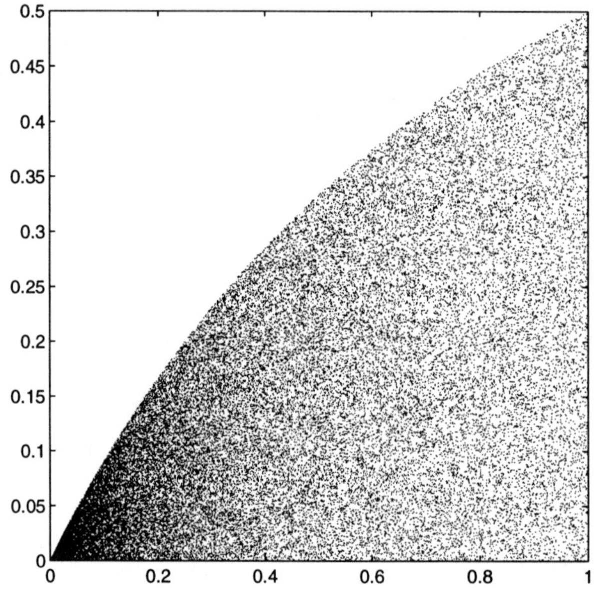

Figure 5.2 *Typical sequence of iterates* (ϵ_n, x_n)*,* $n = 1, \ldots, 50,000$*, in the pessimistic case*

An easier derivation of these densities is by direct use of the transform $(\epsilon, s) \mapsto (\epsilon, x)$. Possession of the invariant densities for (ϵ, x) directly gives the one-dimensional marginal densities for (x_n) derived above and allows computation of characteristics of the process, such as the Lyapunov exponents Λ_1, Λ_2.

These are obtained by integrating the log of the eigenvalues λ_1, λ_2 of the Jacobian with respect to the invariant density. Since the Jacobians are lower triangular, the eigenvalues are the diagonal entries. In the pessimistic case, $\Lambda_1 = -\Lambda_2$, with

$$\Lambda_2 \;=\; \int_0^1 \log\left(1 - \epsilon\left\lfloor\frac{1}{\epsilon}\right\rfloor\right) \frac{1}{(1+\epsilon)\log 2} d\epsilon$$

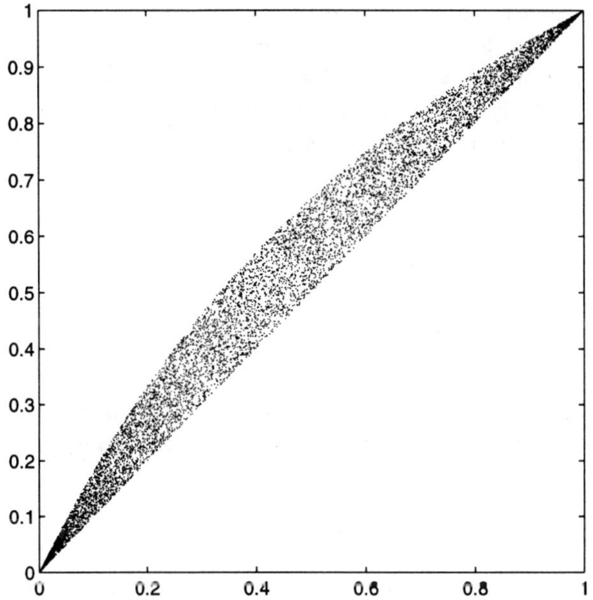

Figure 5.3 *Typical sequence of iterates* (ϵ_n, x_n), $n = 1, \ldots, 10,000$, *in the optimistic case*

$$= \sum_{m=1}^{\infty} \int_{1/(m+1)}^{1/m} \frac{\log(1-\epsilon m)}{(1+\epsilon)\log 2} d\epsilon$$

$$= -\frac{\pi^2}{6\log 2}.$$

In the optimistic case, again $\tilde{\Lambda}_1 = -\tilde{\Lambda}_2$, with

$$\tilde{\Lambda}_2 = \int_0^1 \log\left[\frac{(1-\epsilon\left\lfloor\frac{1}{\epsilon}\right\rfloor)(\epsilon\left\lfloor\frac{1}{\epsilon}\right\rfloor+\epsilon-1)}{\epsilon(1-\epsilon)}\right]\frac{1}{(1+\epsilon)\log 2}d\epsilon$$

$$= \sum_{m=1}^{\infty}\int_{1/(m+1)}^{1/m}\log\left[\frac{(1-\epsilon m)(\epsilon m+\epsilon-1)}{\epsilon(1-\epsilon)}\right]\frac{1}{(1+\epsilon)\log 2}d\epsilon$$

$$= -\frac{\pi^2}{6\log 2}.$$

An important warning is that, although in some cases the Lyapunov exponents correspond directly to the log-rate of the algorithm (see Theorem 4.1), this does *not* apply in this case. The reason is that the initialisation $\epsilon_1 = x^*$ destroys the condition that the state variables $\boldsymbol{\theta}_n$ should not depend explicitly on x^*; see Section 4.1. However, the log-rates can be computed directly using definition (4.3) and the invariant density. For the pessimistic and optimistic cases, the rates are respectively (using the notations in Section 2.2.2)

$$r_n = \frac{L_{n+1}}{L_n} = \frac{q_{n-1}q_n}{q_n q_{n+1}} = s_{n+2}s_{n+1} ,$$

and

$$\tilde{r}_n = \frac{q_n(q_{n-1}+q_n)}{q_{n+1}(q_n+q_{n+1})} = s_{n+2}^2 \frac{1+s_{n+1}}{1+s_{n+2}} .$$

The marginal for s_n is the same in both cases, namely the same as for the Gauss map: $\phi(x) = 1/[(1+x)\log 2]$ for $x \in [0,1]$, which gives

$$\varrho = \frac{\pi^2}{6\log 2} \simeq 2.37314$$

in both cases, twice the Lyapunov exponent of the Gauss map. It is mystifying, considering the warning just given, that the values of $\Lambda_1, \tilde{\Lambda}_1$ for the (ϵ_n, x_n) processes are the same.

As discussed in Section 4.2, W_2 seems to be a more appealing characteristic of the asymptotic precision of the approximation than $W_1 = \varrho$. For finite n and x^* uniform in $[0,1)$,

$$\mathbf{E}L_n = \mathbf{E}\{L_n(x^*)\} = \sum_{i_1,\ldots,i_n} p_{i_1,\ldots,i_n}^2 ,$$

where

$$p_{i_1,\ldots,i_n} = |\{x_1 \in [0,1) \;/\; x_1 \in \mathcal{I}_{i_1}, \ldots, x_n \in \mathcal{I}_{i_n}\}| ,$$

with $\mathcal{I}_i = [1/(i+1), 1/i)$ and each i_k in the summation runs from 1 to ∞. The probabilities $p_{i_1,\ldots,i_n}$ are exactly the lengths of the optimistic intervals after n iterations. Noticing that the i_k's are exactly the partial quotients a_k in the continued fraction expansion of x^*, we get

$$p_{i_1,\ldots,i_n} = \frac{1}{q_n(q_{n-1}+q_n)} ,$$

see (3.7), with $q_n = q_n(a_1, \ldots, a_n)$. Therefore,

$$W_2 = -\lim_{n\to\infty} \frac{1}{n} \log \mathbf{E}L_n$$

$$= -\lim_{n\to\infty}\frac{1}{n}\log\sum_{i_1,\ldots,i_n}\left[\frac{1}{q_n(q_{n-1}+q_n)}\right]^2,$$

and more generally for $\gamma \geq 0$ and $\gamma \neq 1$,

$$\begin{aligned} W_\gamma &= \frac{1}{1-\gamma}\lim_{n\to\infty}\frac{1}{n}\log \mathbf{E}\mathbf{L}_n^{\gamma-1} \\ &= \frac{1}{1-\gamma}\lim_{n\to\infty}\frac{1}{n}\log\sum_{i_1,\ldots,i_n}\left[\frac{1}{q_n(q_{n-1}+q_n)}\right]^\gamma . \end{aligned}$$

Vallée (1997) and Flajolet and Vallée (1998) show that the limits above exist for any $\gamma > 1$ and

$$W_\gamma = \frac{1}{1-\gamma}\log\lambda_\gamma ,$$

where λ_γ is the maximum eigenvalue of the so-called Ruelle-Mayer transfer operator:

$$G_\gamma[f](t) = \sum_{j=1}^{\infty}\frac{1}{(j+t)^{2\gamma}}f\left(\frac{1}{j+t}\right).$$

For $\gamma = 1$, this is just the Frobenius-Perron operator and the invariant density for the Gauss map (3.3) is the eigenfunction corresponding to the (maximum) eigenvalue 1. Numerical values of λ_γ for some γ's are given in Flajolet and Vallée (1998); for instance, for $\gamma = 2$,

$$W_2 \simeq 1.61219 ,$$

which is called the "Vallée constant."

The extreme values of W_γ are easily computed. For $\gamma = 0$, W_0 counts the number of cells in the partitions $\mathcal{Q}_n$; see (8.9), and thus $W_0 = \infty$. For $\gamma = \infty$ one has

$$W_\infty = -\lim_{n\to\infty}\frac{1}{n}\log\max_{a_1,\ldots,a_n}\left\{\frac{1}{q_n(q_{n-1}+q_n)}\right\}.$$

The maximum over $a_1,\ldots,a_n$ is achieved when $a_1 = a_2 = \cdots = a_n = 1$, which gives $q_n = F_{n+1}$, with (F_n) the sequence of Fibonacci numbers; see Section 2.2.3. Therefore,

$$\max_{a_1,\ldots,a_n}\left\{\frac{1}{q_n(q_{n-1}+q_n)}\right\} = \frac{1}{F_{n+1}F_{n+2}} .$$

Using (2.11), we get

$$\begin{aligned} W_\infty &= -\lim_{n\to\infty}\frac{1}{n}\log\frac{5}{(1+\varphi)^{2n+3}} \\ &= 2\log(1+\varphi) \simeq 0.96242\,. \end{aligned}$$

5.3 Golden-Section algorithm

5.3.1 Ergodic behaviour for symmetric functions

In the special case when $f(x)$ is symmetric around x^* the renormalised algorithm yields the time-homogeneous dynamical process defined by (3.14) and shown in Figure 3.8. This dynamical system exhibits a chaotic behaviour and has an invariant measure μ_T such that

$$\lim_{k\mapsto\infty}\frac{1}{k}\sum_{i=1}^{k} g(x_i) = \int_0^1 g(x)\mu_T(dx)$$

for any continuous function $g(\cdot)$ on $[0,1)$ and μ_T-almost all starting points $x_0 \in [0,1]$. The measure μ_T has the density (with respect to the Lebesgue measure on $[0,1)$) $p(x)$ given by

$$p(x) = \begin{cases} 0 & \text{if } x \notin [\frac{1-\varphi}{2}, \frac{1+\varphi}{2}) \\ \frac{4+3\varphi}{5} & \text{if } \frac{1-\varphi}{2} \le x < \frac{\varphi}{2}, \\ \frac{7+4\varphi}{5} & \text{if } \frac{\varphi}{2} \le x < 1-\frac{\varphi}{2} \\ \frac{4+3\varphi}{5} & \text{if } 1-\frac{\varphi}{2} \le x < \frac{1+\varphi}{2} \end{cases} \tag{5.6}$$

and shown on Figure 5.4.

The convergence rate at any iteration $n \ge 1$ is equal to φ, and thus for any $\gamma \ge 0$, $W_\gamma = \varrho = -\log\varphi \simeq 0.4812118$. Most results about the asymptotic behaviour of line-search algorithms hold under the assumption that the function $f(\cdot)$ is uni-extremal and symmetric. In the next section the condition of symmetry will be relaxed, and we shall consider locally symmetric functions satisfying

$$f(x) = f(x^*) + C_1|x-x^*|^\gamma + O(|x-x^*|^{\beta+\gamma}) \tag{5.7}$$

for some $\beta > 0$, with $C_1 > 0, \gamma > 0$. Note that if $f(\cdot)$ is smooth at x^*, then (5.7) holds with $\gamma = 2$. We then show that the asymptotic behaviour of the GS algorithm is the same as for symmetric functions. The same property holds for some other line-search algorithms, based on section-invariant numbers.

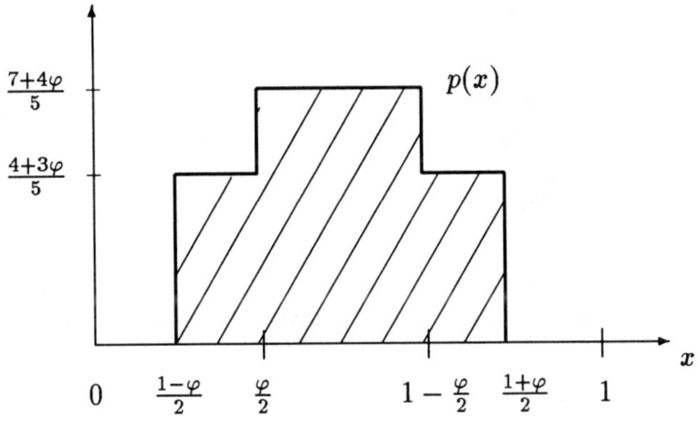

Figure 5.4 *Graph of the invariant density for the GS algorithm*

5.3.2 Extension to locally symmetric functions

In this section the function $f(\cdot)$ will be assumed to be locally symmetric and uni-extremal at x^*. It thus satisfies (5.7), which we rewrite in the following form:

$$C_1|x-x^*|^\gamma - C_2|x-x^*|^{\beta+\gamma} \leq \\ |f(x)-f(x^*)| \leq C_1|x-x^*|^\gamma + C_2|x-x^*|^{\beta+\gamma} \quad (5.8)$$

for any x in $[0,1)$, for some $\beta > 0$, $C_1 > 0$, $C_2 \geq 0$ and $\gamma > 0$. Throughout this section we shall still denote by (x_k) the dynamic process (3.14) obtained for a symmetric function, but shall denote by (x'_k) the process defined by (3.12), which corresponds to the actual function $f(\cdot)$ satisfying (5.8). We thus define the sequence

$$x'_{n+1} = \begin{cases} x'_n(1+\varphi) & \text{if } f_n(1-\varphi) < f_n(\varphi) \quad (\mathrm{R}') \\ x'_n(1+\varphi)-\varphi & \text{if } f_n(1-\varphi) \geq f_n(\varphi) \quad (\mathrm{L}') \end{cases} \quad (5.9)$$

This section aims at proving that the processes (x_n) and (x'_n) have the same asymptotic behaviour.

Before transforming (5.8) to the renormalised function $f_n(\cdot)$, we first derive the expression of $f_n(\cdot)$ for the GS algorithm. Consider the deletion rule (3.14), and define at iteration n

$$\alpha_n = \begin{cases} 0 & \text{if (R)} \\ 1 & \text{if (L)} \end{cases}$$

We then have $x_1 = x^* = \varphi x_2 + \alpha_1(1-\varphi)$, and by induction

$$x^* = g_n^{-1}(x_n) = \sum_{i=1}^{n} \alpha_i(1-\varphi)\varphi^{i-1} + \varphi^n x_n \,,$$

which defines $g_n^{-1}(z)$ and thus $f_n(\cdot) = f(g_n^{-1}(\cdot))$ for any z in $[0,1)$.

Note that the same decision rules (R′) or (L′) are obtained if $f_n(\cdot)$ at iteration n is replaced by

$$\bar{f}_n(\cdot) = K_1(n) f_n(\cdot) + K_2(n) \,,$$

for some $K_1(n), K_2(n)$. Then, $\bar{f}_n(\cdot)$ will be our renormalised function, uni-extremal at x_n, and we have the following lemma.

Lemma 5.1 *If $K_1(n) = (1+\varphi)^{n\gamma}$ and $K_2(n) = -(1+\varphi)^{n\gamma} f(x^*)$ then $\bar{f}_n(\cdot)$ satisfies*

$$\begin{aligned} m_n(x_n, z) = C_1|z - x_n|^\gamma - C_2\varphi^{n\beta} \quad &\le \quad \bar{f}_n(z) \le \\ C_1|z - x_n|^\gamma + C_2\varphi^{n\beta} \quad &= \quad M_n(x_n, z) \,. \end{aligned} \tag{5.10}$$

Proof. The substitution of $g_n^{-1}(z)$ for x in (5.8) and the multiplication by $K_1(n)$ gives

$$\begin{aligned} K_2(n) + (1+\varphi)^{n\gamma} f(x^*) \quad &+ \quad C_1|z-x_n|^\gamma - C_2|z-x_n|^{\gamma+\beta}\varphi^{n\beta} \\ &\le \quad \bar{f}_n(z) \\ \le K_2(n) + (1+\varphi)^{n\gamma} f(x^*) \quad &+ \quad C_1|z-x_n|^\gamma + C_2|z-x_n|^{\gamma+\beta}\varphi^{n\beta} \,. \end{aligned}$$

Since $|z - x_n| \le 1$, and using the definition of $K_2(n)$, we get (5.10). ■

We shall say that

(i) a *mistake* has been made at iteration n of the process (x'_k) if $x_n = x'_n$ and $x_{n+1} \neq x'_{n+1}$;

(ii) the mistake at iteration n has been corrected in m iterations if $x_{n+l} \neq x'_{n+l}$ for $1 < l < m$ and $x_m = x'_m$;

(iii) we pay a penalty $m-1$ for any mistake corrected in m iterations.

The penalty s_N paid up to iteration N counts the number of pairs (x_n, x'_n), $n \le N$, in the sequences (x_n) and (x'_n) such that $x_n \neq x'_n$; that is,

$$s_N = \{\text{number of } n \le N \text{ such that } x_n \neq x'_n\} \,. \tag{5.11}$$

From the definition of the Golden-Section algorithm and the uni-extremality of $f(\cdot)$ no mistake can happen outside the interval $[1-\varphi, \varphi]$. The next statement shows that as n increases, the size of the interval where mistakes can occur decreases exponentially.

Lemma 5.2 *The interval $\mathcal{V}_n$ where a mistake can occur at iteration n is included in*

$$\mathcal{U}_n = [\frac{1}{2} - C_3\varphi^{n\beta}, \frac{1}{2} + C_3\varphi^{n\beta}], \tag{5.12}$$

where

$$C_3 = 2\frac{C_2}{C_1}(3+2\varphi)^\gamma \max\{1, (2\varphi - 1)\frac{2^{\gamma-2}}{\gamma}\}. \tag{5.13}$$

Proof. No mistake is possible at iteration n if

$$M_n(x_n, \varphi) < m_n(x_n, 1-\varphi) \tag{5.14}$$

or

$$m_n(x_n, \varphi) > M_n(x_n, 1-\varphi), \tag{5.15}$$

where $m_n(x_n, z)$ and $M_n(x_n, z)$ are the minorant and majorants of $\bar{f}_n(z)$ defined in (5.10). We consider the case $x_n \in (1/2, \varphi)$ (the case $x_n \in (1-\varphi, 1/2)$ can be treated in the same way). The inequality (5.14) is equivalent to

$$C_1(\varphi - x_n)^\gamma + C_2\varphi^{n\beta} < C_1[x_n - (1-\varphi)]^\gamma - C_2\varphi^{n\beta}.$$

Denote $z = x_n - 1/2$, with $0 < z < \varphi - 1/2$, and rewrite the last inequality in the form

$$\frac{2C_2\varphi^{n\beta}}{C_1} < [(\varphi - \frac{1}{2}) + z]^\gamma - [(\varphi - \frac{1}{2}) - z]^\gamma = \psi_\gamma(z). \tag{5.16}$$

For $1 \le \gamma \le 2$ the function $\psi_\gamma(\cdot)$ is concave, and a sufficient condition for (5.16) to hold is

$$\frac{2C_2\varphi^{n\beta}}{C_1} < [\psi_\gamma(\varphi - \frac{1}{2}) - \psi_\gamma(0)]z = 2^\gamma(\varphi - \frac{1}{2})^\gamma z \le \psi_\gamma(z).$$

For $0 < \gamma \le 1$ and $\gamma \ge 2$ the function $\psi_\gamma(\cdot)$ is convex, and a sufficient condition for (5.16) to hold is now

$$\frac{2C_2\varphi^{n\beta}}{C_1} < \psi_\gamma(0) + \psi'_\gamma(0)z = 2\gamma(\varphi - \frac{1}{2})^{\gamma-1} z \le \psi_\gamma(z).$$

The solution of the left-hand side inequalities with respect to z yields the upper end of $\mathcal{U}_n$ given by (5.12–5.13). ■

Define the intervals

$$\mathcal{A}_k = [\frac{1}{2} - l_k, \frac{1}{2} + l_k] \subset [0, 1),$$

with

$$l_k = (\varphi - \frac{1}{2})(2\varphi - 1)^k = P_k + \varphi Q_k, \quad k = 1, 2, \dots$$

where the generating functions for the coefficients P_k and Q_k are

$$P(z) = -\frac{1 - z}{2(1 - 4z - z^2)}, \quad Q(z) = -\frac{1}{1 - 4z - z^2}.$$

The next lemma states a so-called *self-correcting property* for the GS algorithm: any mistake within the interval $\mathcal{A}_1 = [4\varphi - 2, 3 - 4\varphi]$ will be necessarily corrected in three iterations. Moreover, it shows that for all $k \geq 1$, any point in the set $\mathcal{A}_k \backslash \mathcal{A}_{k+1}$ moves to $\mathcal{A}_{k-1} \backslash \mathcal{A}_k$ in three iterations of (3.14).

Lemma 5.3 Self-correcting property for the Golden-Section algorithm.
For any $n \geq 1$, $k \geq 1$ if $x_n = x'_n \in \mathcal{A}_k \setminus \mathcal{A}_{k+1}$ and $x_{n+1} \neq x'_{n+1}$ then $x_{n+3} = x'_{n+3} \in \mathcal{A}_{k-1} \setminus \mathcal{A}_k$.

Proof. Let

$$x_n = x'_n \in (\frac{1}{2} + l_{k+1}, \frac{1}{2} + l_k]; \tag{5.17}$$

that is, $x_n = x'_n \in \mathcal{A}_k \setminus \mathcal{A}_{k+1} \cap [1/2, 1]$. Applying three iterations of (3.14) we obtain

$$\begin{aligned}
x_{n+1} &= (1+\varphi)x_n - \varphi \\
&\in \left((\frac{1}{2} + l_{k+1})(1+\varphi) - \varphi, (\frac{1}{2} + l_k)(1+\varphi) - \varphi\right], \\
x_{n+2} &= (1+\varphi)x_{n+1} = (2+\varphi)x_n - 1 \\
&\in \left((\frac{1}{2} + l_{k+1})(2+\varphi) - 1, (\frac{1}{2} + l_k)(2+\varphi) - 1\right], \\
x_{n+3} &= (1+\varphi)x_{n+2} = (3+2\varphi)x_n - (1+\varphi) \\
&\in \left((\frac{1}{2} + l_{k+1})(3+2\varphi) - (1+\varphi), (\frac{1}{2} + l_k)(3+2\varphi) - \right. \\
&\qquad \left. (1+\varphi)\right] \\
&= (\frac{1}{2} + l_k, \frac{1}{2} + l_{k-1}],
\end{aligned}$$

where the last inequality follows from the definition of the $\mathcal{A}_k$'s. Assume (5.17) again and $x'_{n+1} \neq x_{n+1}$; that is, a mistake is made at iteration n. Then we get

$$x'_{n+1} = (1+\varphi)x_n \in \left((\frac{1}{2} + l_{k+1})(1+\varphi), (\frac{1}{2} + l_k)(1+\varphi)\right]$$

by application of (R′) in (5.9). Since $(1/2 + l_{k+1})(1+\varphi) > \varphi$, the (L′) rule is necessarily used at iteration $n+1$ of (5.9), and therefore

$$\begin{aligned} x'_{n+2} &= (1+\varphi)x'_{n+1} - \varphi = (2+\varphi)x_n - \varphi \\ &\in \left((\frac{1}{2} + l_{k+1})(2+\varphi) - \varphi, (\frac{1}{2} + l_k)(2+\varphi) - \varphi\right] . \end{aligned}$$

Since $(1/2 + l_{k+1})(2+\varphi) - \varphi > \varphi$, (L′) is necessarily used again at iteration $n+2$ of (5.9), and

$$\begin{aligned} x'_{n+3} &= (1+\varphi)x'_{n+2} - \varphi = (3+2\varphi)x_n - (1+\varphi) = x_{n+3} \\ &\in \left(\frac{1}{2} + l_k, \frac{1}{2} + l_{k-1}\right] . \end{aligned}$$

The case $x_n < 1/2$ can be treated in the same way. ■

Theorem 5.3 *Let $f(\cdot)$ satisfy the condition (5.8). Then the asymptotic behaviours of the dynamical systems (x_n) and (x'_n) defined respectively by (3.14) and (5.9) coincide in the sense that for almost all $x^* \in [0,1)$ and any Borel set $\mathcal{A} \subset [0,1)$*

$$\lim_{N\to\infty} \frac{1}{N} \sum_{n=1}^{N} \mathrm{I}_{\mathcal{A}}(x'_n) = \mu_T(\mathcal{A}) , \qquad (5.18)$$

where μ_T is the invariant measure for (x_n) with the density (5.6).

Proof. Assume that the unknown target $x^* = x_1 = x'_1$ is such that for any Borel set $\mathcal{A} \subset [0,1)$ the limit

$$\lim_{N\to\infty} \frac{1}{N} \sum_{n=1}^{N} \mathrm{I}_{\mathcal{A}}(x_n) \qquad (5.19)$$

for the dynamical system (3.14) exists and equals $\mu_T(\mathcal{A})$. According to classical results in ergodic theory, the Lebesgue measure of the set $\mathcal{S}^* \subset [0,1)$ of such x^* equals 1.

Consider the intervals $\mathcal{U}_n$ of Lemma 5.2. From Lemma 5.3, for any $x_* \in \mathcal{S}^*$ all mistakes are corrected in a finite number of iterations. Let n_0 be such that $\mathcal{U}_{n_0} \subset \mathcal{A}_1$ and the mistakes of all previous iterations have been corrected. Let $\delta > 0$ be a fixed number and $M = M(\delta)$ be the smallest integer such that

$$\mu_T(\mathcal{U}_M) = 2C_3\varphi^{M\beta}\frac{7+4\varphi}{5} \le \frac{\delta}{2},$$

with the measure μ_T associated with (5.6) and C_3 given by (5.13). This gives

$$\begin{aligned} M &= \max\left\{1, 1 + \left\lfloor \frac{1}{\beta\log(1+\varphi)}\left(\log\frac{1}{\delta} + \log C_3 \right.\right.\right. \\ &\quad \left.\left.\left. + \log\frac{4(7+4\varphi)}{5}\right)\right\rfloor\right\}. \end{aligned}$$

Set $N_* = \max\{n_0, M\}$, we shall give an upper bound on the number of mistakes in the process (5.9) by assuming that we make a mistake every time $x_n \in \mathcal{U}_n$. According to Lemma 5.3, we pay a penalty 2 for each mistake, and we therefore obtain for the penalty (5.11) up to iteration $N > N_*$:

$$\begin{aligned} s_N &\le n_0 + 2\{\text{number of } n \le N \text{ such that } x_n \in \mathcal{U}_n\} \\ &= n_0 + 2\sum_{n=n_0}^{N} \mathrm{I}_{\mathcal{U}_n}(x_n) \\ &\le n_0 + 2(N_* - n_0) + 2\sum_{n=N_*}^{N} \mathrm{I}_{\mathcal{U}_{N_*}}(x_n). \end{aligned}$$

Using the inequality above and the fact that (5.19) equals $\mu_T(\mathcal{A})$ for any Borel set $\mathcal{A} \subset [0,1)$, we get

$$\begin{aligned} &\limsup_{N\to\infty} \frac{s_N}{N} \le \\ &\limsup_{N\to\infty} \left[\frac{n_0 + 2(N_* - n_0)}{N} \right. \\ &\left. + \frac{N - N_*}{N}\frac{2}{N - N_*}\sum_{n=N_*}^{N} \mathrm{I}_{\mathcal{U}_{N_*}}(x_n)\right] \le \\ &2\mu_T(\mathcal{U}_{N_*}) \le \delta. \end{aligned}$$

Since δ can be chosen arbitrarily small, we get

$$\lim_{N\to\infty} \frac{s_N}{N} = 0.$$

Finally, for any $x^* \in \mathcal{S}^*$ and any Borel set $\mathcal{A} \subset [0,1)$, we have

$$\left|\frac{1}{N}\sum_{n=1}^{N} \mathrm{I}_{\mathcal{A}}(x'_n) - \mu_T(\mathcal{A})\right| \leq$$
$$\left|\frac{1}{N}\sum_{n=1}^{N} \mathrm{I}_{\mathcal{A}}(x_n) - \mu_T(\mathcal{A})\right| + \left|\frac{1}{N}\sum_{n=1}^{N}[\mathrm{I}_{\mathcal{A}}(x'_n) - \mathrm{I}_{\mathcal{A}}(x_n)]\right|$$
$$\leq \left|\frac{1}{N}\sum_{n=1}^{N} \mathrm{I}_{\mathcal{A}}(x_n) - \mu_T(\mathcal{A})\right| + \frac{s_N}{N} \quad \rightarrow 0 \text{ as } N \rightarrow \infty\,,$$

which gives (5.18). ∎

5.4 Ergodically optimal second-order line-search algorithm

5.4.1 Lower bound for the ergodic convergence rate

Consider the case of minimisation over $[A, B)$, when we know that $f(x)$ is symmetric about x^*. Let u_n, v_n be as in (3.10), $c_n = (u_n + v_n)/2$, and suppose $f(u_n) > f(v_n)$. Then symmetry implies that $x_n > c_n$. We can thus achieve a greater interval reduction than in the usual case by eliminating $[0, c_n)$ rather than only $[0, u_n)$. We refer to this as the *optimistic* rule. The recurrent relation for (x_n) and the rates (r_n) for this rule are given by the formulae

$$x_{n+1} = \begin{cases} \frac{x_n}{c_n} & \text{if } x_n < c_n \\ \frac{x_n - c_n}{1-c_n} & \text{if } x_n \geq c_n \end{cases} \qquad r_n = \begin{cases} c_n & \text{if } x_n < c_n \\ 1 - c_n & \text{if } x_n \geq c_n \end{cases} \tag{5.20}$$

One can clearly achieve the ergodic rate $r = 1/2$ if $r_n = 1/2$ for all n. A way to do this is always to place the new observation point at one of the end points of the new interval obtained by the optimistic rule. An important question is: can we achieve an ergodic rate less than $1/2$ for all x^* lying in a set of positive Lebesgue measure? The answer is *no* as shown below.

Theorem 5.4 *For any sequence* (c_k), $0 < c_k < 1$, $k = 0, 1, \ldots$

$$\mu_{\mathcal{L}}\{x_1 \in [0,1) \;/\; \liminf_{n\to\infty} [L_n(x_1)]^{\frac{1}{n}} < \frac{1}{2}\} = 0\,,$$

where $\mu_{\mathcal{L}}$ *is the Lebesgue measure and* $L_n(x_1) = L_0 r_0 r_1 r_2 \ldots r_{n-1}$, *with* $L_0 = B_0 - A_0$.

Proof. Any second-order minimisation algorithm using the optimistic rule is determined by a sequence (c_n). After n iterations, the interval $\mathcal{K}_n$ for x_1 for which the reduction coefficients coincide with $(r_k)_{k=0}^{n-1}$ has the length $L_n(x_1)$.

Let us fix δ, $0 < \delta < 1/2$, and consider the sets

$$\mathcal{X}_{n,\delta} = \{x_1 \in [0,1) \;/\; [L_n(x_1)]^{\frac{1}{n}} < \frac{1}{2} - \delta\}\,.$$

The total length of the union of the sub-intervals $\mathcal{K}_n \cap \mathcal{X}_{n,\delta}$ cannot exceed the number of disjoint intervals $\mathcal{K}_n$ times the value of L_n and is therefore bounded above by $2^n(1/2-\delta)^n$. Thus, applying Borel-Cantelli arguments

$$\begin{aligned}\mu_{\mathcal{L}}\{x_1 \in [0,1) \;/\; \liminf_{n\to\infty}[L_n(x_1)]^{\frac{1}{n}} < \frac{1}{2} - \delta\} &= \\ \mu_{\mathcal{L}}\{x_1 \in [0,1) \;/\; [L_n(x_1)]^{\frac{1}{n}} < \frac{1}{2} - \delta \text{ infinitely often }\} &= 0\,,\end{aligned}$$

since for any $0 < \delta < 1/2$

$$\sum_{n=1}^{\infty} \mu_{\mathcal{L}}\{x_1 \in [0,1) \;/\; [L_n(x_1)]^{\frac{1}{n}} < \frac{1}{2} - \delta\} \le \sum_{n=1}^{\infty} 2^n(\frac{1}{2} - \delta)^n < \infty\,.$$

This completes the proof. ■

The statement of the theorem implies that there are no second-order minimisation algorithms with an ergodic rate smaller than 1/2 (or a log-rate larger than 2) for the class of symmetric functions, and therefore for any wider class of objective functions.

5.4.2 Achieving the lower bound

We can construct an algorithm which achieves the optimal ergodic rate of 1/2 for a class of locally smooth functions satisfying

$$f(x) = f(x^*) + (x-x^*)^2\,\frac{f''(x^*)}{2} + O(|x-x^*|^{2+\beta}) \quad \text{for some } \beta > 0\,.$$

The algorithm and the full proof of its convergence are very technical and can be found in Wynn and Zhigljavsky (1995). The algorithm spends most of its time behaving in the optimistic fashion defined by (5.20). To avoid getting trapped, a simple test is made which leads either to the continuation of the optimistic process or to a correction involving *backtracking* to the point where a wrong judgement was made. Thus, the algorithm requires an optimistic uncertainty interval $[u'_n, v'_n]$ constructed according to the optimistic rule which lies inside the ordinary (renormalised) interval $[0,1)$. When $v'_n - u'_n$ becomes smaller than a predefined value δ_n, a check is made outside $[u'_n, v'_n]$ to test for a mistake. Backtracking then switches to a new correct optimistic interval. It is shown in Wynn and Zhigljavsky (1995) that the parameter δ_n may be controlled in such a way that when x^* is a normal number to the base 2, see Section 3.2.1, the number of corrections is finite and the effect of all iterations "wasted" on checking is asymptotically negligible.

A family of ϵ-optimal algorithms, based on section-invariant numbers, is presented in Section 5.7.2 (Example 5.10).

5.5 Symmetric algorithms

Throughout this section we consider renormalisation to the closed interval $[0, 1]$.

5.5.1 Relation with continued fractions

There is a most interesting connection between the discussion in Section 5.2 and second-order line-search algorithms such as GS.

We consider a general symmetric algorithm. This is a primitive generalisation of the GS algorithm as follows. Using the notation of Section 3.2.3, we define a symmetric algorithm by

$$\begin{aligned} u_n &= \min(e_n, e'_n), \\ v_n &= \max(e_n, e'_n), \\ e_{n+1} &= \begin{cases} \frac{u_n}{v_n} & (R) \\ \frac{v_n - u_n}{1-u_n} & (L) \end{cases} \end{aligned} \tag{5.21}$$

and

$$e'_n = 1 - e_n$$

for every n. This last condition implies $u_n = 1 - v_n$, which gives both for the (R) and (L) cases $r_n = v_n$ and

$$r_{n+1} = \begin{cases} \frac{1-r_n}{r_n} & \text{if } 1/2 \le r_n < 2/3 \\ 2 - \frac{1}{r_n} & \text{if } 2/3 \le r_n \le 1 \end{cases}$$

with (r_n) living in $[1/2, 1]$. Note that the updating formula for r_n implies by induction that, for fixed n, the function $L_n(E_1)/L_1$ in Figure 2.3 is piecewise linear.

The connection with the Gauss and Farey maps stems from the behaviour of (r_n), which does not depend on the objective function on which the line-search algorithm is applied. No assumption of symmetry of the function is thus required to study this behaviour. Figure 5.5 presents a plot of a typical sequence of iterates (x_n, x_{n+1}), $n = 1, \ldots, 100,000$ for a non-periodic symmetric algorithm, with x_n the renormalised location of x^*.

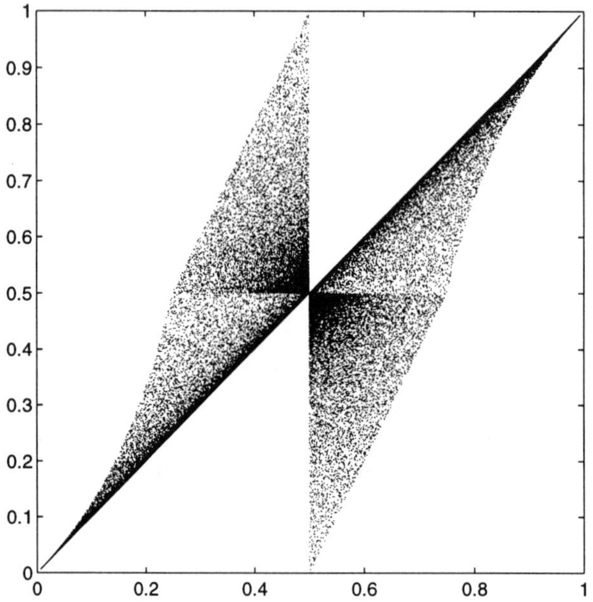

Figure 5.5 *Typical chaotic sequence of iterates* (x_n, x_{n+1}), $n = 1, \ldots, 100,000$, *in a symmetric algorithm*

The simple transformation $z_n = (1/r_n) - 1$ gives the Farey map; that is, the evolution of z_n corresponds to the dynamical system defined by (3.8), the critical point 2/3 for r_n being transformed to 1/2. Note that any statement on the asymptotic behaviour of (z_n) deriving from the Farey map which translates into the behaviour of the continued fraction approximation (see Section 3.2.2) now is mapped into the behaviour of (r_n), which is the same for almost all initial values e_1. This implies for example that the invariant density for (r_n) is $\phi(r) = 1/[r(1-r)]$. Note that the behaviour (and the convergence rate) of the symmetric algorithm is completely determined by e_1. For instance the best convergence rate is for the GS algorithm; that is, when $e_1 = \varphi$, the Golden Section, as shown at the end of the section.

Another way to represent the behaviour of a symmetric second-order line-search algorithm through the Farey map is as follows. As was already discussed, (r_n), the sequence of rates in the symmetric algorithm, depends on the initial value e_1 and not on the objective function. Therefore, (r_n) is independent of whether we apply (L) or (R) in the iterations of (5.21). We can thus assume for simplicity that the objective function monotonously increases, so that the rule (R) always applies. Then, the evolution of (e_n) becomes

$$e_{n+1} = \frac{u_n}{v_n} = \begin{cases} \frac{e_n}{1-e_n} & \text{if } 0 \le e_n < \frac{1}{2} \\ \frac{1-e_n}{e_n} & \text{if } \frac{1}{2} \le e_n \le 1 \end{cases} \tag{5.22}$$

which is exactly the Farey map. The associated sequence of rates is

$$r_n = \begin{cases} 1 - e_n & \text{if } 0 \le e_n < \frac{1}{2} \\ e_n & \text{if } \frac{1}{2} \le e_n \le 1 \end{cases} \tag{5.23}$$

5.5.2 L_n as a function of e_1

The following theorem gives the expression of the length of the uncertainty interval $[A_n, B_n]$ for a symmetric algorithm as a function of the continued fraction expansion of e_1. The expression is valid for any objective function on which the line-search algorithm is applied.

Theorem 5.5 *Consider a symmetric second-order line-search algorithm, initialised at e_1 with continued fraction expansion $e_1 = [a_1, a_2, \ldots]$. Define*

$$n_0 = 0\,,\; n_j = \sum_{i=1}^{j} a_i\,,\; j \ge 1\,. \tag{5.24}$$

Then, for any N such that $n_j \le N < n_{j+1}$, $j \ge 0$,

$$L_{N+1} = L_1 \times \epsilon_0 \times \cdots \times \epsilon_j \times [1 - (N - n_j)\epsilon_{j+1}]\,, \tag{5.25}$$

where $\epsilon_0 = 1$, $\epsilon_1 = e_1$ and $\epsilon_{j+1} = \{1/\epsilon_j\}$, $j \ge 1$.

Proof. We assume that $e_1 < 1/2$. The case $e_1 > 1/2$ could be treated similarly. Recalling the relation between the Farey and continued fraction maps discussed in Section 3.2.2, we notice that between two successive visits to the interval $[1/2, 1]$, say the j-th and $(j+1)$-th, the sequence (5.22) spends $a_{j+1} - 1$ iterations in $[0, 1/2)$. The subsequence (e_{n_j}), with n_j given by (5.24), then contains all terms of the sequence (e_n) that belong to $[1/2, 1)$. Moreover, for

$j \geq 0$ e_{n_j+1} exactly coincides with ϵ_{j+1} in the Gauss-map sequence (ϵ_j), defined by $\epsilon_{j+1} = \{1/\epsilon_j\}$ and $\epsilon_1 = e_1$. Also, if we consider $N = n_j$ iterations of the algorithm (5.22), then we arrive at the point $e_{n_j+1} = \epsilon_{j+1}$ and the length of the unnormalised uncertainty interval after these N iterations is equal to

$$L_{n_j+1} = L_1 \times r_1 \times \cdots \times r_{n_j} = L_1 \times \epsilon_1 \times \cdots \times \epsilon_j \,. \tag{5.26}$$

This can be proved as follows.

Let $j = 0$, and compute the reduction rate in n_1 iterations, from $e_{n_0+1} = e_1 = \epsilon_1$ to $e_{n_1+1} = \epsilon_2$. The length of the uncertainty interval then becomes

$$L_{n_1+1} = L_{n_0+1} \times r_{n_0+1} \times \cdots \times r_{n_1} \,.$$

Since

$$\begin{cases} e_i < \frac{1}{2} & \text{and} \quad r_i = 1 - e_i \,,\ i = n_0+1, \ldots, n_1 - 1 \,, \\ \frac{1}{2} \leq e_{n_1} & \text{and} \quad r_{n_1} = e_{n_1} \end{cases}$$

one gets by induction

$$e_i = \frac{e_1}{1-(i-1)e_1} \,,\ r_i = \frac{1 - ie_1}{1-(i-1)e_1} \,,\ i = n_0+1, \ldots, n_1 - 1 \,, \tag{5.27}$$

and $r_{n_1} = e_{n_1} = e_{n_0+1}/[1-(n_1-1)e_{n_0+1}]$. Therefore,

$$L_{n_1+1} = L_{n_0+1}e_1 = L_{n_0+1}\epsilon_1 \,.$$

The same arguments generalise to arbitrary $j \geq 1$, and $L_{n_j+1} = L_{n_{j-1}+1}\epsilon_j$, which gives (5.26). Finally, using (5.27), one can easily get the value of L_{N+1} for arbitrary N; that is, (5.25) for $n_j \leq N < n_{j+1}$. ■

Theorem 5.5 implies in particular that, when e_1 is a rational in $[0, 1)$, the length of the uncertainty interval remains constant after some iteration. Indeed, one can write $e_1 = [a_1, \ldots, a_m, \infty]$; see (2.4). This gives $\epsilon_{m+1} = 0$, $n_{m+1} = \infty$, and thus from (5.25)

$$L_{N+1} = L_1 \times \epsilon_1 \times \cdots \times \epsilon_m$$

for any $N \geq N^* = a_1 + \cdots + a_m$. Thus, for rational e_1, the algorithm terminates (degenerates) at a certain iteration N^*, where $N^* = a_1 + \cdots + a_m$ could also be defined as the first n such that $e_n = 1$. (Recall that in the last iteration we make the two last observations at the same point, the midpoint of the interval.)

Since the termination point N^* is a function of e_1 we need to be careful in the specification of the sample size of the algorithm.

If we set $N > N^*$ then there is no further improvement after N^* so that the per iteration rate will have declined. However, we have the option of deciding in advance that the number of iterations is $N = N^*$ and then controlling e_1 to produce $\min_{e_1} L_N(e_1)$, the minimal length of the uncertainty interval in N iterations. Fortunately, the case $N = N^*$ gives a nice expression for the value of $L_N(e_1)$ in terms of the so-called *Farey tree*; see definitions below and Corollary 5.1, and the value of e_1 where the minimum $\min_{e_1} L_N(e_1)$ is achieved (see Corollary 5.2).

Consider the mediant $\mathrm{med}(p/q, p'/q')$ of two fractions p/q and p'/q', $\mathrm{med}(p/q, p'/q') = (p+p')/(q+q')$, where p, p', q, q' are positive integers. One can easily check that for $p/q < p'/q'$, $\mathrm{med}(p/q, p'/q')$ always belongs to the interval $(p/q, p'/q')$. The Farey tree (also called *Brocot sequence*, see Lagarias (1992)) $\mathcal{F}_n$ of order n is defined inductively as follows. $\mathcal{F}_0$ consists of two elements, 0 and 1, written as 0/1 and 1/1. Then at iteration n, for every pair $\{p/q, p'/q'\}$ of adjacent fractions in $\mathcal{F}_{n-1}$, their mediant $\mathrm{med}(p/q, p'/q')$ is added to the elements of $\mathcal{F}_{n-1}$. Thus,

$$\mathcal{F}_n = \mathcal{F}_{n-1} \bigcup \mathrm{med}(p/q, p'/q'),$$

where the union is taken over all adjacent pairs $\{p/q, p'/q'\}$ in $\mathcal{F}_{n-1}$. For example,

$$\begin{aligned} \mathcal{F}_1 &= \{0, \frac{1}{2}, 1\}, \quad \mathcal{F}_2 = \{0, \frac{1}{3}, \frac{1}{2}, \frac{2}{3}, 1\}, \\ \mathcal{F}_3 &= \{0, \frac{1}{4}, \frac{1}{3}, \frac{2}{5}, \frac{1}{2}, \frac{3}{5}, \frac{2}{3}, \frac{3}{4}, 1\}. \end{aligned}$$

One can easily check (see Schroeder (1991), p. 337 and Cornfeld, Fomin and Sinai (1982)) that there are exactly 2^{n-1} Farey fractions of level n, that is, elements in $\mathcal{F}_n \setminus \mathcal{F}_{n-1}$, and all of them have a finite continued fraction representation $p/q = [a_1, a_2, \ldots, a_k]$ satisfying $a_1 + \ldots + a_k = n$. Also, the number of elements in $\mathcal{F}_n$ is $|\mathcal{F}_n| = 2^n + 1$ and these elements, apart from 0 and 1, have a continued fraction representation $[a_1, a_2, \ldots, a_k]$ satisfying $a_1 + \ldots + a_k \leq n$.

This gives a characterisation of rational starting points e_1 such that the corresponding symmetric line-search algorithm stops at iteration n. Specifically, these points are exactly the points with a finite continued fraction representation $e_1 = [a_1, a_2, \ldots, a_k]$ with $a_1 + \ldots + a_k = n$ and therefore are exactly the Farey fractions of level n.

If we return to Figure 2.3 and consider the function $L_n(e_1)$ for a given n, we observe that it is piecewise linear, with 2^{n-1} local minima at the Farey fractions of level n and $2^{n-1}+1$ local maxima at $0, 1$ and the Farey fractions of all levels smaller than n.

The following property gives the value of L_N when the algorithm degenerates.

Corollary 5.1 *Let $[A_1, B_1] = [0, 1]$ and $e_1 \in [0, 1]$ be a rational number $e_1 = p/q$ with g.c.d.$(p, q) = 1$ and continued fraction expansion $e_1 = [a_1, a_2, \ldots, a_k]$, $\sum_{i=1}^{k} a_i = N^*$. Then $L_N(e_1) = 1/q$ for all $N \geq N^*$.*

Proof. Consider the sequence $\varepsilon_1, \varepsilon_2, \ldots$ with $\varepsilon_1 = e_1 = p/q$. Then

$$\varepsilon_2 = \frac{1}{\varepsilon_1} - a_1 = \frac{q}{p} - a_1 = \frac{q - pa_1}{p} = \frac{p_2}{p_1},$$

where $p_1 = p$ and $p_2 = q - pa_1$ is some integer such that $p_2 < p_1$ and g.c.d.$(p_1, p_2) = 1$ (the fact that $p_2 < p_1$ is equivalent to the fact that $\varepsilon_2 < 1$).

Analogously, for all $i \leq k$, $\varepsilon_{i-1} = p_i/p_{i-1}$ implies $\varepsilon_i = p_{i+1}/p_i$ with $p_{i+1} < p_i$ and g.c.d.$(p_{i+1}, p_i) = 1$. When $i = k-1$, $\varepsilon_{k-1} = 1/a_k$ so that $p_k = 1$ and $p_{k-1} = a_k$.

According to (5.25), ε_{k-1} is the last term we need in the product $\prod_i \varepsilon_i$ for all $N \geq N^*$ (since $a_{k+1} = \infty$ and $\varepsilon_k = 0$). The product itself equals

$$L_{N+1}(e_1) = \prod_{i=1}^{k-1} \varepsilon_i = \frac{p_1}{q} \frac{p_2}{p_1} \frac{p_3}{p_2} \cdots \frac{p_{k-1}}{p_{k-2}} \frac{1}{p_{k-1}} = \frac{1}{q}, \quad N \geq N^*.$$

■

The Fibonacci algorithm can easily be shown to be optimal among symmetric algorithms.

Corollary 5.2 *(Optimality of the Fibonacci search)*
For a fixed N, the minimum $\min_{e_1} L_N(e_1)$ equals $1/F_{N+1}$ and is achieved at $e_1 = F_N/F_{N+1}$ or $e_1 = F_{N-1}/F_{N+1}$, where $F_1, F_2, \ldots$ is the Fibonacci sequence.

Proof. Without any loss of generality we restrict our attention to the case $e_1 > 1/2$. As already noticed in Section 5.5.1, the function $L_N(e_1)$ is piecewise linear on $[1/2, 1]$; see Figure 2.3. It takes its minimum and maximum values at the rational $e_1 = p/q$ with g.c.d$(p, q) = 1$ and continued fraction expansion $e_1 = [a_1, a_2, \ldots, a_k]$ such that $\sum_{i=1}^{k} a_i \leq N$.

According to Corollary 5.1, $L_N(e_1) = 1/q$ at these points. However, one of the properties of the Farey tree is that the maximum value of the denominator q among all the points p/q with g.c.d$(p,q) = 1$ and continued fraction expansion $p/q = [a_1, a_2, \ldots, a_k]$, such that $\sum_{i=1}^{k} a_i \leq N$, is achieved when $p/q = F_N/F_{N+1}$. This property can easily be proved by induction; see also Schroeder (1991), p. 339. ■

5.5.3 Asymptotic performance

Using (5.25), we can easily evaluate the performance of any symmetric algorithm in terms of the ergodic convergence rate R given by (4.2). Since $L_1 < \infty$ it can be written as

$$R(e_1) = \limsup_{N\to\infty} [\prod_{n=1}^{N} r_n(e_1)]^{\frac{1}{N}} .$$

In what follows we assume, without any loss of generality, that $L_1 = 1$. Next corollary shows how the asymptotic rate of convergence of the algorithm is related to the choice of e_1.

Corollary 5.3 *The convergence of a symmetric algorithm initialised at $e_1 \in (0,1)$ is*
(i) sub-exponential if e_1 is a rational number (in this case, the algorithm is such that the two test points E_n, E'_n coincide at some iteration n),
(ii) exponential if e_1 is a badly approximable number, *that is, a number with bounded partial quotients,*
(iii) sub-exponential for almost all values of e_1.

Proof.

(i) We have seen in Section 5.5.2 that, when e_1 is rational, $L_n(e_1)$ remains constant for n larger than some N, and therefore $R(e_1) = 1$.

(ii) When e_1 is as a badly approximable number, one has $e_1 = [1, a_1, a_2, \ldots]$ with $a_i < A$ for all i and some $A < \infty$. (Note that unlike the set of quadratic irrationals, the set of badly approximable numbers is uncountable. However, it still has zero Lebesgue measure.) From (5.26), we get

$$R(e_1) = \limsup_{j\to\infty} (\epsilon_1 \times \cdots \times \epsilon_j)^{1/n_j} ,$$

with n_j defined by (5.24). Therefore,

$$R(e_1) \leq \limsup_{j\to\infty} (\epsilon_1 \times \cdots \times \epsilon_j)^{1/(jA)} .$$

From the recurrence $\epsilon_{i+1} = \{1/\epsilon_i\}$, we get $\epsilon_i\epsilon_{i+1} < \varphi$. Indeed, the result is obvious for $\epsilon_i \leq \varphi$, and $\epsilon_i > \varphi$ implies $\epsilon_{i+1} = 1/\epsilon_i - 1 < \varphi$. Therefore,

$$R(e_1) \leq \varphi^{1/(2A)} < 1$$

and the convergence is exponential; see Section 4.1 (note that here R does not depend on x^*).

(iii) Ergodic arguments can be used to show that for almost all values of e_1 the convergence is sub-exponential. Indeed, for almost all e_1, the dynamical system (5.22) has the invariant density of the Farey map:

$$\phi_e(x) = \frac{1}{x} \,, \quad 0 < x < 1 \,;$$

see Section 3.2.2. Using (5.23), the density of the sequence (r_n) is easily obtained as

$$\phi_r(x) = \frac{1}{x(1-x)} \,, \quad \frac{1}{2} \leq x < 1 \,.$$

These densities are not normalised; however, they can be used to construct the proportion of points falling in different intervals. Consider in particular the interval $\mathcal{I}_\delta = [1-\delta, 1)$, $0 < \delta < 1/2$, and let n_δ denote the number of points r_i, $i = 1\ldots, n$, in $\mathcal{I}_\delta$. Then n_δ/n tends to 1 for any δ. This gives

$$\begin{aligned} R(e_1) &= \limsup_{n\to\infty} (r_1 \times \cdots \times r_n)^{1/n} \\ &> \limsup_{n\to\infty} \left[(1-\delta)^{n_\delta}(1/2)^{n-n_\delta}\right]^{1/n} = 1-\delta \end{aligned}$$

for any δ. Therefore $R(e_1) = 1$ for almost all e_1. ■

As a complement to the case (iii) above, notice that ergodic properties of the Gauss map imply (see Cornfeld, Fomin and Sinai (1982)) that the following limit

$$\rho'(e_1) = \lim_{j\to\infty} \frac{1}{j} \sum_{i=1}^{j} \log \epsilon_i$$

exists for almost all e_1 in $(0,1)$ and equals

$$\rho' = \rho'(e_1) = \frac{\pi^2}{12 \log 2} \simeq 1.186569 \,,$$

which is the Lyapunov exponent for the Gauss map. A consequence of this relation is that for almost all e_1 in $(0,1)$ for large N

$$\prod_{i=1}^{j} \epsilon_i \sim \exp(j\rho')$$

which together with (5.25) implies $L_N \sim \exp(j\rho')$ where j is such that $n_j \leq N < n_{j+1}$.

Corollary 5.3 (i) does not mean that symmetric algorithms have necessarily poor performance for finite N; see, for instance, the Fibonacci algorithm, for which $e_1 = F_N/F_{N+1} = [1,\ldots,1]$ with N 1's, where (F_N) is the Fibonacci sequence.

A particular case of Corollary 5.3 (ii) is when e_1 is a *quadratic irrational*, that is, solution of a quadratic equation with integer coefficients. Its continued fraction expansion is then periodic starting with some n (and only quadratic irrationals have this property; see Rockett and Szüsz, 1992). Let the period be $b_1,\ldots,b_k$; that is,

$$e_1 = [1, a_1, \ldots, a_n, \underbrace{b_1, \ldots, b_k}, \underbrace{b_1, \ldots, b_k}, \ldots].$$

Then, (5.26) implies that the ergodic rate of convergence of the corresponding algorithm is

$$R(e_1) = (\epsilon_1' \times \cdots \epsilon_k')^{1/(b_1+\cdots+b_k)},$$

where

$$\begin{aligned}
\epsilon_1' &= [\underbrace{b_1, \ldots, b_k}, \underbrace{b_1, \ldots, b_k}, \ldots] \\
\epsilon_2' &= [\underbrace{b_2, \ldots, b_k, b_1}, \underbrace{b_2, \ldots, b_k, b_1}, \ldots] \\
&\;\;\vdots \\
\epsilon_k' &= [\underbrace{b_k, b_1, \ldots, b_{k-1}}, \underbrace{b_k, b_1, \ldots, b_{k-1}}, \ldots]
\end{aligned}$$

All the ϵ_i''s are smaller than 1 and the convergence is therefore exponential.

A famous example is the GS algorithm, for which $e_1 = \varphi = [1, 1, \ldots]$, which gives $R(e_1) = \varphi \simeq 0.61803$. More generally, the same value for R is obtained when e_1 is a so-called *noble number*; that is, when e_1 has a continued fraction expansion $e_1 =$

$[a_1, \ldots, a_n, 1, 1, \ldots]$ ending in all 1's (see Schroeder (1991), p. 392). Some other examples are given in Table 5.5.

e_1	$R(e_1)$
$\varphi = [1, 1, \ldots]$	$\varphi \simeq 0.61803$
$2 - \sqrt{2} = [1, 1, 2, 2, 2, \ldots]$	$\sqrt{\sqrt{2} - 1} \simeq 0.64359$
$\sqrt{2}/2 = [1, 2, 2, 2, \ldots]$	$\sqrt{\sqrt{2} - 1} \simeq 0.64359$
$\sqrt{3} - 1 = [\underbrace{1, 2}, \underbrace{1, 2}, \ldots]$	$(\sqrt{3} - 2)^{1/3} \simeq 0.64469$
$\sqrt{3}/3 = [1, \underbrace{1, 2}, \underbrace{1, 2}, \ldots]$	$(\sqrt{3} - 2)^{1/3} \simeq 0.64469$
$\sqrt{10}/2 - 1 = [\underbrace{1, 1, 2}, \underbrace{1, 1, 2}, \ldots]$	$(\sqrt{10} - 3)^{1/4} \simeq 0.63469$
$(\sqrt{10} - 1)/3 = [\underbrace{1, 2, 1}, \underbrace{1, 2, 1}, \ldots]$	$(\sqrt{10} - 3)^{1/4} \simeq 0.63469$

Table 5.5 *Values of $R(e_1)$ for different choices of quadratic irrationals e_1*

Figure 5.6 presents a plot of a typical sequence of iterates (x_n, r_n), $n = 1, \ldots, 100,000$ for a non-periodic symmetric algorithm, corresponding to Corollary 5.3 (iii). This propery has the important consequence that a direct implementation of a symmetric algorithm, based on the application of the rule $E'_n = A_n + B_n - E_n$, yields sub-exponential convergence ($R = 1$) due to numerical inaccuracies. In particular, this is the case for the GS algorithm, hence the usual recommendation to use the implementation (2.9). Note that among symmetric algorithms, $R(e_1)$ is minimum when $e_1 = \varphi$, the Golden Section. This follows from Corollary 5.2 and the fact that

$$\frac{L_n(\varphi)}{L_n(F_n/F_{n+1})} \to \frac{2 + \varphi}{\sqrt{5}}, \quad n \to \infty;$$

see (2.11).

5.6 Midpoint and window algorithms

5.6.1 Midpoint algorithm

The midpoint algorithm, see Section 2.2.3, always places the new point at $e'_n = 1/2$ and, when $f(\cdot)$ is symmetric about x^*, yields the

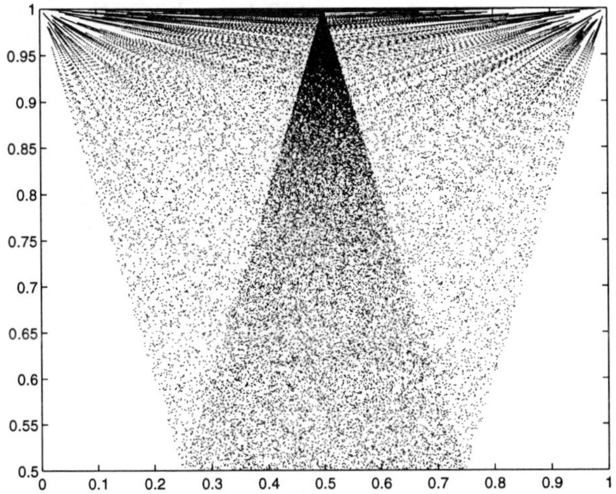

Figure 5.6 *Typical chaotic sequence of iterates* (x_n, r_n), $n = 1, \ldots, 100,000$, *in a symmetric algorithm*

dynamical system

$$x_{n+1} = \begin{cases} 2x_n & \text{if } x_n < c_n, e_n < \frac{1}{2} \\ \frac{x_n}{2e_n} & \text{if } x_n < c_n, e_n \geq \frac{1}{2} \\ 2x_n - 1 & \text{if } x_n \geq c_n, e_n \geq \frac{1}{2} \\ \frac{x_n - e_n}{1 - e_n} & \text{if } x_n \geq c_n, e_n < \frac{1}{2} \end{cases}$$

$$e_{n+1} = \begin{cases} 2e_n & \text{if } x_n < c_n, e_n < \frac{1}{2} \\ \frac{1}{2e_n} & \text{if } x_n < c_n, e_n \geq \frac{1}{2} \\ 2e_n - 1 & \text{if } x_n \geq c_n, e_n \geq \frac{1}{2} \\ \frac{\frac{1}{2} - e_n}{1 - e_n} & \text{if } x_n \geq c_n, e_n < \frac{1}{2} \end{cases}$$

where $c_n = (e_n + e_n')/2 = 1/4 + e_n/2$, $x_1 = x^*$ and e_1 is any irrational number in $[0, 1)$. The dynamical system is two-dimensional, with the renormalised target x^* as one component and the renormalised location of the observation point as the other. The second component e_{n+1} only depends functionally on x_n through the test for left or right deletion, which is in agreement with the discussion in Section 3.1.1, and Theorem 4.1 applies. One can check that the mapping $T^2(., .) = T(T(., .))$ is uniformly expanding, which implies the existence of an invariant measure absolutely continuous with respect to the Lebesgue measure.

Figure 5.7 (respectively 5.8) presents a plot of the sequence of iterates (x_n, e_n) (respectively (x_n, x_{n+1})), $n = 1, \ldots, 50,000$.

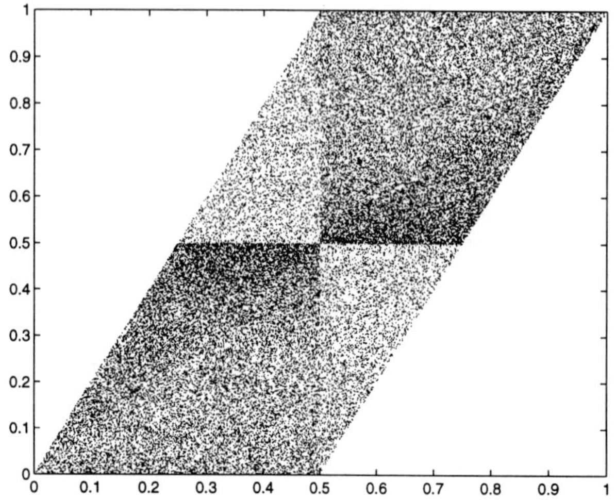

Figure 5.7 *Typical sequence of iterates* (x_n, e_n), $n = 1, \ldots, 50,000$, *in the midpoint algorithm*

The rate associated with the dynamical system (x_n, e_n) satisfies

$$r_n = \begin{cases} \frac{1}{2} & \text{if } x_n < c_n, e_n < \frac{1}{2} \\ e_n & \text{if } x_n < c_n, e_n \geq \frac{1}{2} \\ \frac{1}{2} & \text{if } x_n \geq c_n, e_n \geq \frac{1}{2} \\ 1 - e_n & \text{if } x_n \geq c_n, e_n < \frac{1}{2} \end{cases}$$

The Jacobian matrix of the transformation

$$T : (x_n, e_n) \mapsto (x_{n+1}, e_{n+1})$$

is upper triangular. Its value at the point (x_n, e_n) can easily be computed:

$$\boldsymbol{J}_T(x_n, e_n) = \begin{pmatrix} \frac{1}{r_n} & \alpha_n \\ 0 & \pm\frac{1}{2r_n^2} \end{pmatrix},$$

where α_n is some number. The Lyapunov exponents are

$$\Lambda_1 \quad = \quad - \lim_{N \to \infty} \frac{1}{N} \sum_{n=1}^{N} \log r_n = \varrho \simeq 0.5365\,,$$

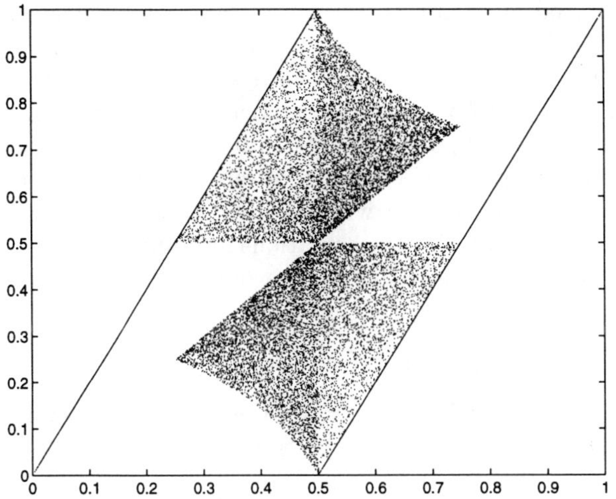

Figure 5.8 *Typical sequence of iterates* (x_n, x_{n+1}), $n = 1, \ldots, 50,000$, *in the midpoint algorithm*

$$\Lambda_2 \quad = \quad -\lim_{N\to\infty} \frac{1}{N} \sum_{n=1}^{N} \log(2r_n^2) = 2\Lambda_1 - \log 2 \simeq 0.3799\,.$$

The numerical values were obtained by simulation as well as by numerical solution of the Frobenius-Perron equation. In agreement with Theorem 4.1, the largest Lyapunov exponent Λ_1 coincides with the log-rate $\varrho = -\log R$. The ergodic rate $R \simeq 0.5848$ of the midpoint algorithm is a little better than the rate $R = \varphi \simeq 0.6180$ of the GS algorithm. Other algorithms exist with even faster ergodic rates.

5.6.2 Window algorithm and fractals

We shall consider here the asymptotic and finite sample behaviour of the window algorithm, with tuning parameters ϵ and w, presented in Section 2.2.3. The algorithm is defined by (2.13, 2.14).

The ergodic characteristic R does not depend on ϵ. Unfortunately, good tuning of w for the ergodic criterion R, for example $w = 1/8$, is far from being optimal for small n for all criteria considered here, whatever the choice of ϵ. Also, for a fixed value of w, choosing ϵ large enough ($\epsilon \geq (1-w)/[2(1+w)]$), that is, expanding the initial interval $[A_0, B_0)$, guarantees that x_1 belongs to the support of the invariant measure for x_n. This is of crucial importance, since it allows us to obtain finite sample characteristics close to their asymptotic values. Also, numerical investigations demonstrate that starting points x_1 outside the support of the invariant measure for x_n give bad convergence rates in the first iterations.

Consider the special case when $f(\cdot)$ is symmetric around x^*. Due to the symmetry of $f(\cdot)$, the conditions for application of the right or left deletion rule in (3.11) then become

$$\begin{cases} (R) & \text{if } x_n < e_n + w/2 \\ (L) & \text{if } x_n \geq e_n + w/2 \,. \end{cases}$$

Renormalisation back to $[0, 1)$ yields the following two-dimensional dynamical process

$$\begin{aligned} x_{n+1} &= \begin{cases} \frac{x_n}{e_n+w} & \text{if } x_n < e_n + w/2 \\ \frac{x_n-e_n}{1-e_n} & \text{if } x_n \geq e_n + w/2 \end{cases} \\ e_{n+1} &= \begin{cases} \frac{e_n}{e_n+w} - w & \text{if } x_n < e_n + w/2 \\ \frac{w}{1-e_n} & \text{if } x_n \geq e_n + w/2 \end{cases} \end{aligned} \tag{5.28}$$

Figure 5.9 (respectively 5.10) presents a plot of the sequence of iterates (x_n, e_n) (respectively (x_n, x_{n+1})), $n = 1, \ldots, 50,000$, when $w = 1/8$.

The rate at iteration n is

$$r_n = \begin{cases} e_n + w & \text{if } x_n < e_n + w/2 \\ 1 - e_n & \text{if } x_n \geq e_n + w/2 \end{cases}$$

with $r_0 = 1 + 2\epsilon$.

The Jacobian matrix of the transformation

$$T : (x_n, e_n) \mapsto (x_{n+1}, e_{n+1})$$

is upper triangular and can be written as

$$\boldsymbol{J}_T(x_n, e_n) = \begin{pmatrix} \frac{1}{r_n} & \beta_n \\ 0 & \frac{w}{r_n^2} \end{pmatrix},$$

where β_n is some number. This shows that the Lyapunov exponents of the dynamical system are related to R and w by $\Lambda_1 = \varrho = -\log R$, $\Lambda_2 = 2\Lambda_1 + \log w$. Numerical simulations show that the second Lyapunov exponent is typically negative, which indicates that the dynamical system (x_n, u_n) may attract to a fractal. For instance when $w = 1/8$, the Lyapunov exponents, determined by

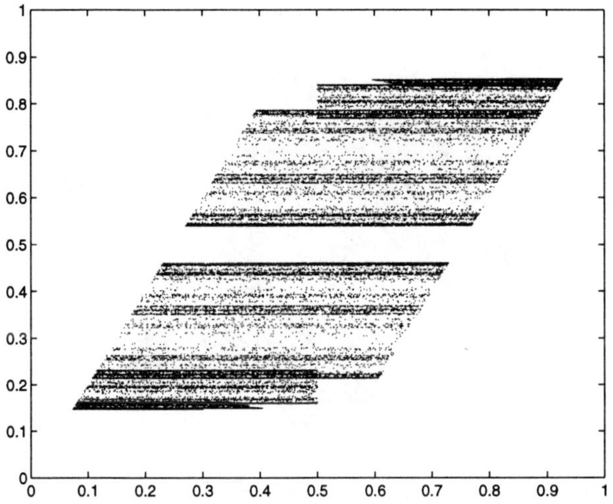

Figure 5.9 *Typical sequence of iterates* (x_n, e_n), $n = 1, \ldots, 50{,}000$, *in the window algorithm with* $w = 1/8$

numerical simulations, are $\Lambda_1 \simeq 0.639$, $\Lambda_2 \simeq -0.801$. The Lyapunov dimension of the attractor of the corresponding dynamical system, presented on Figure 5.9, is then $1 - \Lambda_1/\Lambda_2 \simeq 1.798 < 2$. The ergodic rate is $R \simeq 0.528$. When $w = 0.15$, the value that will be used in Section 5.8, we get $\Lambda_1 \simeq 0.630$, $\Lambda_2 \simeq -0.636$ and $R \simeq 0.532$. This increase in the value of R permits, however, obtaining better performances for finite N ($N \leq 30$). More generally, for any w in $[1/8, 2\varphi - 1)$ the ergodic rate of the window algorithm is better than the rate $R = \varphi = (\sqrt{5} - 1)/2$ of the GS algorithm.

5.7 Algorithms based on section-invariant numbers

In all this section $f(\cdot)$ is taken to be symmetric about x^*.

5.7.1 Dynamical system representation

Consider the evolution of x_n, the renormalised value of x^* in $[0, 1)$ for an algorithm based on section-invariant numbers; see Section 2.2.3. Let $\mathcal{U} = \{v_1, \ldots, v_m\}$ be a collection of section-invariant

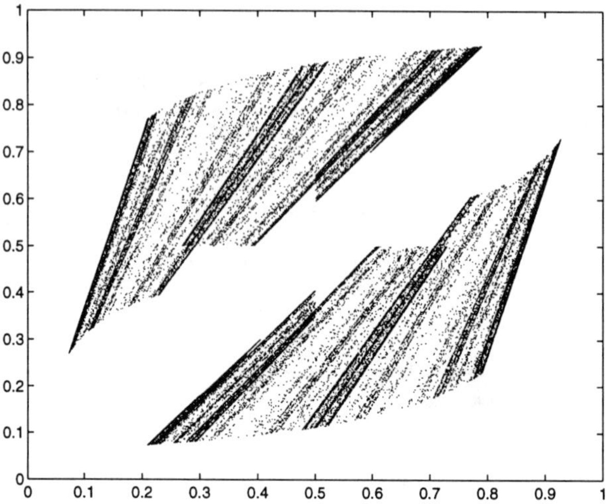

Figure 5.10 *Typical sequence of iterates* (x_n, x_{n+1}), $n = 1, \ldots, 50,000$, *in the window algorithm with* $w = 1/8$

numbers. The dynamical system is then defined by

$$x_{n+1} = \begin{cases} \frac{x_n}{v_n} & \text{if } x_n < \frac{u_n+v_n}{2} \quad (R) \\ \frac{x_n-u_n}{1-u_n} & \text{otherwise} \quad (L) \end{cases}$$

where $u_n = \min\{e_n, e'_n\}$, $v_n = \max\{e_n, e'_n\}$, and $e_n = v_{j_n} \in \mathcal{U}$, $e'_n = v'_{j_n} \in \mathcal{U}'$, for some $j_n \in \{1, \ldots, m\}$. Then, $e'_{n+1} = R(v_{j_n}, v'_{j_n})$ when (R) applies and $e'_{n+1} = L(v_{j_n}, v'_{j_n})$ when (L) applies; see (2.12). The process is then completely described by the trajectory of (j_n, x_n) which evolves in $\{1, \ldots, m\} \times [0, 1)$ and is defined by

$$(j, x) \to \begin{cases} (R(j), \frac{x}{v_j}) & \text{if } x < c_j \\ (L(j), \frac{x-u_j}{1-u_j}) & \text{otherwise} \end{cases} \tag{5.29}$$

where $u_j = \min\{v_j, v'_j\}$, $v_j = \max\{v_j, v'_j\}$, $R(j)$ (resp. $L(j)$) is such that $u_{R(j)} = R(v_j, v'_j)$ (resp. $u_{L(j)} = L(v_j, v'_j)$), and

$$c_j = \frac{v_j + v'_j}{2}. \tag{5.30}$$

In order to study the ergodic properties of the process, we shall define an equivalent dynamical process in $[0, 1)$. The interval $[0, 1)$ is divided into m sub-intervals $\Delta_j = [(j-1)/m, j/m)$, $j = 1, \ldots, m$, with v_j associated with Δ_j. With each pair (j, x) we also associate a new variable z,

$$z = \frac{x}{m} + \frac{j-1}{m}. \tag{5.31}$$

The corresponding dynamical process is $z_{n+1} = T(z_n)$, where

$$\begin{aligned} T(z) &= \sum_{j=1}^{m} \Big(\frac{z - j/m}{1-u_j} + \frac{L(j)}{m}\Big) \mathtt{I}_{[c_j/m+(j-1)/m, j/m)}(z) \\ &+ \sum_{j=1}^{m} \Big(\frac{z-(j-1)/m}{v_j} + \frac{R(j)-1}{m}\Big) \\ &\quad \times \mathtt{I}_{[(j-1)/m, c_j/m+(j-1)/m)}(z). \end{aligned} \tag{5.32}$$

Note that in (5.32), the two indicators $\mathtt{I}_{[c_j/m+(j-1)/j, j/m)}(z)$ and $\mathtt{I}_{[(j-1)/m, c_j/m+(j-1)/m)}(z)$ respectively correspond to the conditions $x \geq c_j$, $x < c_j$ in (5.29). Also note that the convention used to define the mapping at the points j/m, $j = 1, \ldots, m$, is arbitrary since these values of z can only be reached if the process is initialised at $x_1 = 0$ or 1, and in this case $x_n = 0$ or 1 for all n. The process z_n associated with a given algorithm is thus characterised by a piecewise linear mapping $T : [0,1) \mapsto [0,1)$. The behaviour of such systems has been studied in Section 4.5. We have seen in particular that the invariant density for z is

$$\phi_z(z) = \sum_{i=1}^{M} \phi_i \mathtt{I}_{\mathcal{I}_i}(z),$$

with $\phi_i = \bar{\pi}_i / |\mathcal{I}_i|$. From the density for z_n we deduce the density for x_n. Consider the transformation (5.31) and define $\phi_x^{(j)}(\cdot)$ as follows:

$$\phi_x^{(j)}(\cdot) = \mathtt{Pr}(j_n = j) \times \phi_x(\cdot | j_n = j),$$

that is, the conditional density of x when $j_n = j$ multiplied by the invariant probability of being in this state. We have

$$\phi_x^{(j)}(x) = \frac{1}{m} \phi_z \left(\frac{x + (j-1)}{m} \right),$$

and

$$\phi_x(x) = \sum_{i=1}^{N} \phi_x^{(i)}(x) = \frac{1}{N} \sum_{i=1}^{N} \phi_z \left(\frac{x + (i-1)}{N} \right).$$

Example 5.4 *Take $\mathcal{U} = \{1/2, 2/3\}$, $\mathcal{U}' = \{3/4, 1/3\}$; see Table 8.1(b).*

The mapping $T(\cdot)$ for the process z_n is presented in Figure 5.11.

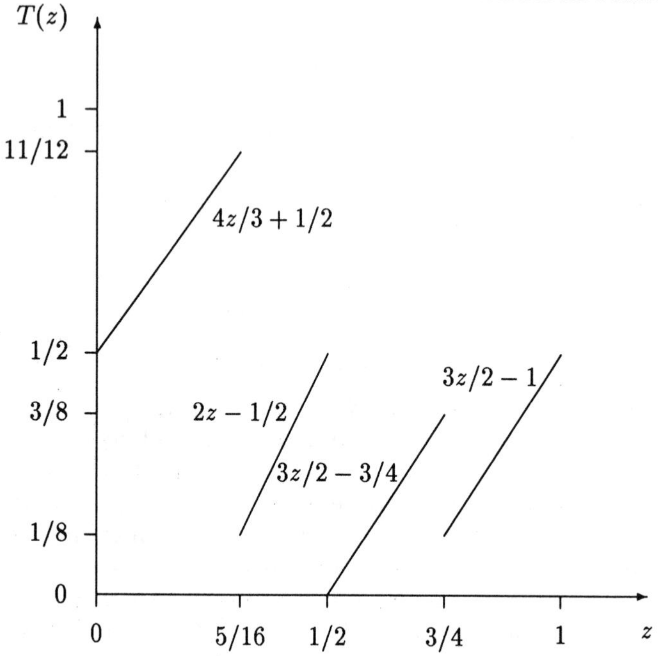

Figure 5.11 *Graph of the mapping* $z \to T(z)$ *when* $\mathcal{U} = \{1/2, 2/3\}$, $\mathcal{U}' = \{3/4, 1/3\}$

The sets $\mathcal{A}$ *and* $\mathcal{S}^\infty$ *are*

$$\begin{aligned} \mathcal{A} &= \{0, 5/16, 1/2, 3/4, 1\}, \\ \mathcal{S}^\infty &= \{0, 1/8, 1/4, 5/16, 3/8, 1/2, 2/3, 3/4, 5/6, 11/12, 1\}. \end{aligned}$$

The intervals $\mathcal{I}_i$ *that define the partition of* $[0,1]$ *are*

$$\begin{array}{lll} \mathcal{I}_1 = [0, 1/8], & \mathcal{I}_2 = [1/8, 1/4], & \mathcal{I}_3 = [1/4, 5/16], \\ \mathcal{I}_4 = [5/16, 3/8], & \mathcal{I}_5 = [3/8, 1/2], & \mathcal{I}_6 = [1/2, 2/3], \\ \mathcal{I}_7 = [2/3, 3/4], & \mathcal{I}_8 = [3/4, 5/6], & \mathcal{I}_9 = [5/6, 11/12], \\ \mathcal{I}_{10} = [11/12, 1]. & & \end{array}$$

The mapping $T(\cdot)$ *is then defined as follows:*

$$\begin{array}{l} \mathcal{I}_1 \mapsto \mathcal{I}_6\,,\ \mathcal{I}_2 \mapsto \mathcal{I}_7 \cup \mathcal{I}_8\,,\ \mathcal{I}_3 \mapsto \mathcal{I}_9\,,\ \mathcal{I}_4 \mapsto \mathcal{I}_2\,, \\ \mathcal{I}_5 \mapsto \mathcal{I}_3 \cup \mathcal{I}_4 \cup \mathcal{I}_5\,,\ \mathcal{I}_6 \mapsto \mathcal{I}_1 \cup \mathcal{I}_2\,,\ \mathcal{I}_7 \mapsto \mathcal{I}_3 \cup \mathcal{I}_4\,, \end{array}$$

$$\mathcal{I}_8 \mapsto \mathcal{I}_2\,,\ \mathcal{I}_9 \mapsto \mathcal{I}_3 \cup \mathcal{I}_4\,,\ \mathcal{I}_{10} \mapsto \mathcal{I}_5\,.$$

The matrix $\boldsymbol{\Pi}$ *is thus*

$$\boldsymbol{\Pi} = \begin{pmatrix} 0 & 0 & 0 & 0 & 0 & 1 & 0 & 0 & 0 & 0 \\ 0 & 0 & 0 & 0 & 0 & 0 & 1/2 & 1/2 & 0 & 0 \\ 0 & 0 & 0 & 0 & 0 & 0 & 0 & 0 & 1 & 0 \\ 0 & 1 & 0 & 0 & 0 & 0 & 0 & 0 & 0 & 0 \\ 0 & 0 & 1/4 & 1/4 & 1/2 & 0 & 0 & 0 & 0 & 0 \\ 1/2 & 1/2 & 0 & 0 & 0 & 0 & 0 & 0 & 0 & 0 \\ 0 & 0 & 1/2 & 1/2 & 0 & 0 & 0 & 0 & 0 & 0 \\ 0 & 1 & 0 & 0 & 0 & 0 & 0 & 0 & 0 & 0 \\ 0 & 0 & 1/2 & 1/2 & 0 & 0 & 0 & 0 & 0 & 0 \\ 0 & 0 & 0 & 0 & 1 & 0 & 0 & 0 & 0 & 0 \end{pmatrix},$$

from which we determine

$$\bar{\boldsymbol{\pi}} = (0, 2/7, 1/7, 1/7, 0, 0, 1/7, 1/7, 1/7, 0)^T\,,$$

and the density ϕ_x*:*

$$\phi_x(x) = \frac{8}{7}\mathrm{I}_{[\frac{1}{4},\frac{3}{4}]}(x) + \frac{6}{7}\mathrm{I}_{[\frac{1}{3},\frac{5}{6}]}(x)\,.$$

Consider now the GS4 algorithm defined in Section 2.2.3. The updating rule for (x_n, e_n) is

$$(x_{n+1}, e_{n+1}) = \begin{cases} (\frac{x_n}{a'}, c) & \text{if } e_n = a \text{ and } x_n < \frac{a+a'}{2} \\ (\frac{x_n - a}{1-a}, a) & \text{if } e_n = a \text{ and } x_n \geq \frac{a+a'}{2} \\ (\frac{x_n}{c}, d) & \text{if } (e_n = b \text{ or } e_n = c) \\ & \text{and } x_n < \frac{b+c}{2} \\ (\frac{x_n - b}{c}, a) & \text{if } (e_n = b \text{ or } e_n = c) \\ & \text{and } x_n \geq \frac{b+c}{2} \\ (\frac{x_n}{d}, d) & \text{if } e_n = d \text{ and } x_n < 1 - \frac{a+a'}{2} \\ (\frac{x_n-(1-a')}{a'}, b) & \text{if } e_n = d \text{ and } x_n \geq 1 - \frac{a+a'}{2} \end{cases} \tag{5.33}$$

where

$$\begin{aligned} b &= 2a^3 - 4a^2 + 3a\,,\ c = 1 - b\,,\ d = 1 - a\,,\ a' = 2a - a^2 \text{ and} \\ a &\simeq 0.19412 \end{aligned} \tag{5.34}$$

is the smallest positive root of the polynomial $2t^4 - 8t^3 + 11t^2 - 7t + 1$.

Due to the symmetry with respect to 1/2 of the possible values of v_j in $\mathcal{U}$ and the symmetry of the possible choices of v' in $\mathcal{U}'$, the dynamical system can be simplified as follows. Define for $n \geq 2$

$$(y_n, g_n) = \begin{cases} (x_n, e_n) & \text{if } e_n = a \text{ or } e_n = b \\ (1 - x_n, 1 - e_n) & \text{if } e_n = c \text{ or } e_n = d \end{cases}$$

Since the rate of convergence of the algorithm is the same for the initial values x_1 and $1 - x_1$, we define

$$(y_1, g_1) = \begin{cases} (x_1, b) & \text{if } x_1 < \frac{1}{2} \\ (1 - x_1, b) & \text{if } x_1 \geq \frac{1}{2} \end{cases} \tag{5.35}$$

so that $y_1 \leq 1/2$. Then, the new system (y_n, g_n) obeys the simplified updating rule

$$(y_{n+1}, g_{n+1}) = \begin{cases} (1 - \frac{y_n}{a'}, b) & \text{if } g_n = a \text{ and } y_n < \frac{a+a'}{2} \\ (\frac{y_n - a}{1-a}, a) & \text{if } g_n = a \text{ and } y_n \geq \frac{a+a'}{2} \\ (1 - \frac{y_n}{c}, a) & \text{if } g_n = b \text{ and } y_n < \frac{1}{2} \\ (\frac{y_n - b}{c}, a) & \text{if } g_n = b \text{ and } y_n \geq \frac{1}{2} \end{cases} \tag{5.36}$$

The price for this simplification is that knowing (y_n, g_n) we do not know whether (x_n, e_n) equals $(y_n, g_n), (1 - y_n, g_n), (y_n, 1 - g_n)$ or $(1-y_n, 1-g_n)$. However, this has no consequence on the calculation of the performance characteristics of Chapter 4. In order to obtain a one-dimensional dynamical system on $[0, 1]$, we now introduce

$$\begin{cases} z_n = \frac{y_n}{2} & \text{if } g_n = a \\ z_n = \frac{1+y_n}{2} & \text{if } g_n = b \end{cases}$$

which gives

$$z_{n+1} = T(z_n) = \begin{cases} 1 - \frac{z_n}{a'} & \text{if } z_n < \frac{a+a'}{4} \\ \frac{2z_n - a}{2(1-a)} & \text{if } \frac{a+a'}{4} \leq z_n < \frac{1}{2} \\ \frac{1}{2} - \frac{2z_n - 1}{2c} & \text{if } \frac{1}{2} \leq z_n < \frac{3}{4} \\ \frac{2z_n - 1 - b}{2c} & \text{if } \frac{3}{4} \leq z_n \end{cases} \tag{5.37}$$

The transformation $T(\cdot)$ is presented on Figure 5.12. Note that the convention used to define the mapping $T(\cdot)$ at $1/2$ is arbitrary since $1/2$ can be reached only if the process is initialised at $x_1 = A_1$ or B_1, which is impossible if $[A_1, B_1]$ is taken strictly larger than $[A_0, B_0]$; see Section 2.2.3.

The convergence rate at iteration n is

$$r_n = \begin{cases} a' & \text{if } z_n < \frac{a+a'}{4} \\ 1 - a & \text{if } \frac{a+a'}{4} \leq z_n < \frac{1}{2} \\ c & \text{if } \frac{1}{2} \leq z_n \end{cases}$$

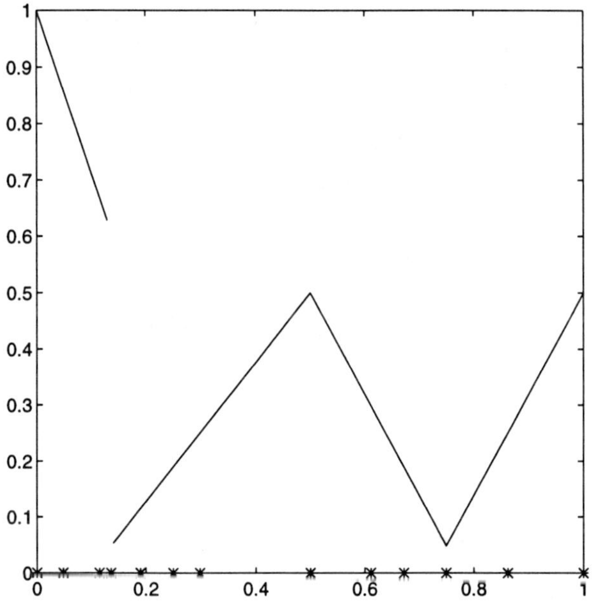

Figure 5.12 *Graph of the mapping of the transformation $T(\cdot)$ for the GS4 algorithm*

Define $\mathcal{A}$ as the collection of all points mentioned in the right-hand side of (5.37); that is, $\mathcal{A} = \{0, \frac{a+a'}{4}, \frac{1}{2}, \frac{3}{4}, 1\}$. Define $\mathcal{S}^\infty = \cup_{n=1}^\infty T^n(\mathcal{A})$. Straightforward calculations, using computer algebra, show that when a is the smallest positive root of $2t^4 - 8t^3 + 11t^2 - 7t + 1$, $\mathcal{S}^\infty$ is finite and given by $\mathcal{S}^\infty = \{\alpha_0, \alpha_1, \ldots, \alpha_{12}\}$, with the α_i's given in Table 5.6, which correspond to the stars in Figure 5.12.

The set $\mathcal{S}^\infty$ defines a partition of $[0,1]$ into 12 intervals $\mathcal{I}_i = [\alpha_{i-1}, \alpha_i], i = 1, \ldots, 12$.

Assume that z_1 has a probability density ϕ_z^1 on $[0,1]$, such that $\phi_z^1(z)$ is constant on each interval $\mathcal{I}_i, i = 1, \ldots, 12$. In particular, when x^* is uniformly distributed on $[A_0, B_0]$:

$$\begin{aligned} e_1 &= b\,, \text{ so that } \phi_z^1(z) = 0\,,\ z \in [0, \frac{1}{2}] = \cup_{i=1}^7 \mathcal{I}_i\,, \\ \epsilon &= \frac{1-a}{2}\,, \text{ so that } \phi_z^1(z) = 0\,,\ z \in \mathcal{I}_8\,. \end{aligned}$$

α_0	0	0
α_1	$\frac{a}{4}$	$\simeq 0.04853$
α_2	$\frac{a^3}{2} - 2a^2 + \frac{9}{4}a - \frac{1}{4}$	$\simeq 0.11506$
α_3	$-\frac{a^2}{4} + \frac{3}{4}a$	$\simeq 0.13617$
α_4	$\frac{a^3}{2} - \frac{3}{2}a^2 + \frac{5}{4}a$	$\simeq 0.18978$
α_5	$\frac{1}{4}$	0.25
α_6	$\frac{1}{4} + \frac{a}{4}$	$\simeq 0.29853$
α_7	$\frac{1}{2}$	0.5
α_8	$\frac{a^3}{2} - a^2 + \frac{3}{4}a + \frac{1}{2}$	$\simeq 0.61156$
α_9	$\frac{a^2}{2} - \frac{a}{2} + \frac{3}{4}$	$\simeq 0.67178$
α_{10}	$\frac{3}{4}$	0.75
α_{11}	$\frac{a^3}{2} - a^2 + \frac{3}{4}a + \frac{3}{4}$	$\simeq 0.86156$
α_{12}	1	1

Table 5.6 $\mathcal{S}^\infty$ *for the GS4 algorithm*

Moreover, $\phi_z^1(z) = 0$, $z \in \mathcal{I}_{11} \cup \mathcal{I}_{12}$ since $y_1 \leq 1/2$ by definition; see (5.35). The density induced on $\mathcal{I}_9$ and $\mathcal{I}_{10}$ is thus

$$\phi_z^1(z) = \frac{1}{\alpha_{10} - \alpha_8} = 8 - 4a \simeq 7.22353\,, \quad z \in \mathcal{I}_9 \cup \mathcal{I}_{10}\,.$$

Similarly, one obtains for the GS4$_0$ algorithm, for which $\epsilon = 0$,

$$\phi_z^1(z) = 4\,, \quad z \in \left[\frac{1}{2}, \frac{3}{4}\right] = \mathcal{I}_8 \cup \mathcal{I}_9 \cup \mathcal{I}_{10}\,,$$

and $\phi_z^1(z) = 0$ for $z \in \mathcal{I}_i$, $i \neq 8, 9, 10$. The density of z_n then remains constant on each interval $\mathcal{I}_i$. Let $\pi_i^{(n)}$ denote the probability $\Pr(z_n \in \mathcal{I}_i)$. The initial distribution, given by $\pi_i^{(1)}, i = 1, \ldots, 12$, is that induced on the states by the prior distribution of x^*, and depends on the choice of ϵ. For the GS4 algorithm, $\epsilon = (1-a)/2$, and

$$\begin{aligned}\boldsymbol{\pi}^{(1)} &= (0,0,0,0,0,0,0,0,-2a^3+6a^2-4a+1,\\ &\quad\; 2a^3-6a^2+4a,0,0)^T \\ &\simeq (0,0,0,0,0,0,0,0,0.4350,0.5650,0,0)^T\,,\end{aligned}$$

while for the GS4$_0$ algorithm $\epsilon = 0$ and

$$\boldsymbol{\pi}^{(1)} = (0,0,0,0,0,0,0,2a^3-4a^2+3a,-2a^3+6a^2-5a+1,$$

$$
\begin{aligned}
&-2a^2 + 2a, 0, 0)^T \\
\simeq\ &(0,0,0,0,0,0,0,0.4462,0.2409,0.3129,0,0)^T .
\end{aligned}
$$

The transition matrix $\boldsymbol{\Pi}$ with elements π_{ij} is given by:

$$
\begin{aligned}
\pi_{1,12} &= 1\,, \quad \pi_{2,10} = -\frac{2}{5}a^3 + \frac{2}{5}a^2 + \frac{2}{5}\,, \\
\pi_{2,11} &= \frac{2}{5}a^3 - \frac{2}{5}a^2 + \frac{3}{5}\,, \quad \pi_{3,9} = 1\,, \\
\pi_{4,2} &= 1\,, \quad \pi_{5,3} = a^3 - 3a^2 + 2a\,, \\
\pi_{5,4} &= -a^3 + 3a^2 - 2a + 1\,, \quad \pi_{6,5} = 1\,, \\
\pi_{7,6} &= a\,, \quad \pi_{7,7} = 1 - a\,, \quad \pi_{8,7} = 1\,, \\
\pi_{9,5} &= -2a^3 + 4a^2 - 3a + 1\,, \quad \pi_{9,6} = 2a^3 - 4a^2 + 3a\,, \\
\pi_{10,2} &= -a^3 + 2a^2 - \frac{a}{2} + \frac{1}{2}\,, \quad \pi_{10,3} = a^2 - 2a + \frac{1}{2}\,, \\
\pi_{10,4} &= a^3 - 3a^2 + \frac{5}{2}a\,, \quad \pi_{11,2} = -2a^3 + 4a^2 + a\,, \\
\pi_{11,3} &= 2a^2 - 5a + 1\,, \quad \pi_{11,4} = 2a^3 - 8a^2 + 8a - 1\,, \\
\pi_{11,5} &= 2a^2 - 4a + 1\,, \quad \pi_{12,6} = a\,, \quad \pi_{12,7} = 1 - a\,,
\end{aligned}
$$

the other elements being equal to 0. This gives the following matrix, where a dot denotes 0.

$$
\boldsymbol{\Pi} \simeq
\begin{pmatrix}
\cdot & \cdot & \cdot & \cdot & \cdot & \cdot & \cdot & \cdot & \cdot & \cdot & \cdot & 1 \\
\cdot & \cdot & \cdot & \cdot & \cdot & \cdot & \cdot & \cdot & \cdot & .412 & .588 & \cdot \\
\cdot & \cdot & \cdot & \cdot & \cdot & \cdot & \cdot & \cdot & 1 & \cdot & \cdot & \cdot \\
\cdot & 1 & \cdot & \cdot & \cdot & \cdot & \cdot & \cdot & \cdot & \cdot & \cdot & \cdot \\
\cdot & \cdot & .283 & .717 & \cdot & \cdot & \cdot & \cdot & \cdot & \cdot & \cdot & \cdot \\
\cdot & \cdot & \cdot & \cdot & 1 & \cdot & \cdot & \cdot & \cdot & \cdot & \cdot & \cdot \\
\cdot & \cdot & \cdot & \cdot & \cdot & .194 & .806 & \cdot & \cdot & \cdot & \cdot & \cdot \\
\cdot & \cdot & \cdot & \cdot & \cdot & \cdot & 1 & \cdot & \cdot & \cdot & \cdot & \cdot \\
\cdot & \cdot & \cdot & \cdot & .554 & .446 & \cdot & \cdot & \cdot & \cdot & \cdot & \cdot \\
\cdot & .471 & .149 & .380 & \cdot & \cdot & \cdot & \cdot & \cdot & \cdot & \cdot & \cdot \\
\cdot & .330 & .105 & .266 & .299 & \cdot & \cdot & \cdot & \cdot & \cdot & \cdot & \cdot \\
\cdot & \cdot & \cdot & \cdot & \cdot & .194 & .806 & \cdot & \cdot & \cdot & \cdot & \cdot
\end{pmatrix}
$$

From (4.17), the vector $\boldsymbol{r}$ of rates associated with the twelve states is

$$
\boldsymbol{r} = (a', a', a', 1-a, 1-a, 1-a, 1-a, c, c, c, c, c)^T . \qquad (5.38)
$$

The invariant distribution $\bar{\pi}$ satisfies

$$\bar{\pi} = \boldsymbol{\Pi}^T \bar{\pi} .$$

This means that $\bar{\pi}$ is given by the normalised eigenvector associated with the eigenvalue 1 for the matrix $\boldsymbol{\Pi}^T$. For the GS4 and $GS4_0$ algorithms, the distribution $\bar{\pi}$ is given in Table 5.7.

$\bar{\pi}_1$	0	
$\bar{\pi}_2$	$\frac{1}{1763}(112a^3 + 4a^2 - 879a + 657)$	$\simeq 0.2764$
$\bar{\pi}_3$	$\frac{1}{1763}(100a^3 - 752a^2 + 1293a - 106)$	$\simeq 0.0666$
$\bar{\pi}_4$	$\frac{1}{1763}(-792a^3 + 2994a^2 - 2977a + 769)$	$\simeq 0.1691$
$\bar{\pi}_5$	$\frac{1}{1763}(-276a^3 + 242a^2 + 592a + 81)$	$\simeq 0.1152$
$\bar{\pi}_6$	$\frac{1}{1763}(644a^3 - 1740a^2 + 1557a - 189)$	$\simeq 0.0297$
$\bar{\pi}_7$	0	
$\bar{\pi}_8$	0	
$\bar{\pi}_9$	$\frac{1}{1763}(100a^3 - 752a^2 + 1293a - 106)$	$\simeq 0.0666$
$\bar{\pi}_{10}$	$\frac{1}{1763}(244a^3 - 495a^2 - 89a + 235)$	$\simeq 0.1139$
$\bar{\pi}_{11}$	$\frac{1}{1763}(-132a^3 + 499a^2 - 790a + 422)$	$\simeq 0.1625$
$\bar{\pi}_{12}$	0	

Table 5.7 *Invariant distribution for the Markov chain in the GS4 and $GS4_0$ algorithms*

In many cases, as in the examples above (Example 5.4 and GS4 algorithm), M is finite, and $T(\cdot)$ becomes a Markov map. The computation of invariant distributions is then straightforward. However, the fact that partitions $\mathcal{S}^\infty$ might be infinite can be proved by an example.

Example 5.5 *Consider Table 8.5, and let c be any non-algebraic number, $1/2 < c < \varphi$. Then, the finiteness of $\mathcal{S}^\infty$ would imply that c satisfies some polynomial equation.*

5.7.2 Performance characteristics

The performance characteristics of algorithms that can be represented by piecewise linear mappings have been studied in Section 4.5. We consider first a collection of examples.

Example 5.6 *Take again $\mathcal{U} = \{1/2, 2/3\}$, $\mathcal{U}' = \{3/4, 1/3\}$; see Table 8.1(b). From (4.17), with the slope of the transformation*

$T(\cdot)$ read on Figure 5.11, we get the ergodic rates $\varrho = \frac{4}{7}\log 2$ and $R = \frac{19}{28}$. The value of W_2 is $-\log\lambda_{\max}$, with $\lambda_{\max}$ the largest real root of $16t^4 - 8t^2 - t + 1$, that is, $\lambda_{\max} \simeq 0.6748$, and $W_2 \simeq 0.3933$.

Example 5.7 *(GS algorithm)*

There are two states only, with $\mathcal{U} = \{v_1, v_2\} = \{1-\varphi, \varphi\}$, and $v_1' = v_2$, $v_2' = v_1$, and the dynamical process is defined by (3.14). The invariant measure has the step-function density shown in Figure 5.4.

The asymptotic rates are $R = \varphi \simeq 0.6180$ and $\varrho = W_1 = -\log\varphi \simeq 0.4812$. The Golden-Section algorithm is the best among those with $|\mathcal{U}| = 2$. The zeta function (4.25) equals

$$\zeta(t) = \frac{1}{1 - 2t + 2t^4 - t^6},$$

which gives the following values for the numbers of periodic trajectories of period n, see (4.24), for $n = 1, \ldots, 10$: 2, 4, 8, 8, 12, 22, 30, 48, 80, 124.

As we shall see soon, when $|\mathcal{U}| > 2$ there are algorithms that are better than the GS algorithm (see also Pronzato, Wynn and Zhigljavsky, 1997, where several algorithms with $|\mathcal{U}| = 5$ and $|\mathcal{U}| = 6$ are detailed).

Example 5.8 *The example corresponds to Table 8.4 The invariant density for x is given by:*

$$\phi_x(x) = \phi_x^{(1)}(x) + \phi_x^{(2)}(x) + \phi_x^{(3)}(x),$$

with

$$\begin{aligned}
\phi_x^{(1)}(x) &= K\mathbb{I}_{[\frac{a}{2},\frac{1}{2}]}(x), \\
\phi_x^{(2)}(x) &= K[2(\psi^2+1)\mathbb{I}_{[\frac{b}{2},\frac{a+b}{2}]}(x) + (\psi^2+\psi+1)\mathbb{I}_{[\frac{a+b}{2},\frac{1}{2}]}(x) \\
&\quad +(\psi^2-\psi+2)\mathbb{I}_{[\frac{1}{2},\frac{1+b}{2}]}(x)], \\
\phi_x^{(3)}(x) &= K[2(1-\psi+\psi^2)\mathbb{I}_{[\frac{c}{2},\frac{1}{2}]}(x) + 2(\psi^2+1)\mathbb{I}_{[\frac{1}{2},\frac{b+c}{2}]}(x) \\
&\quad +(2\psi^2+\psi+1)\mathbb{I}_{[\frac{b+c}{2},\frac{\psi+c}{2}]}(x) \\
&\quad +(3-\psi+2\psi^2)\mathbb{I}_{[\frac{\psi+c}{2}\frac{\psi+1}{2}]}(x) \\
&\quad +(\psi^2-\psi+3)\mathbb{I}_{[\frac{\psi+1}{2},\frac{c+1}{2}]}(x)],
\end{aligned}$$

and the asymptotic rates are given by

$$\varrho \;=\; -\frac{K}{2}[(4\psi^2 - 3\psi + 3)\log\psi + (3\psi^2 - 3\psi + \log(1-\psi)$$

$$+(5\psi^2 - 11\psi + 6)\log(1-\psi+\psi^2)] \simeq 0.5575\,,$$

$$R = \frac{K}{2}(5\psi^2 + 6\psi - 2) = \frac{1017\psi^2 - 1529\psi + 1753}{2075} \simeq 0.5841\,,$$

where ψ is defined as in Table 8.1(d) and

$$K = \frac{2}{12\psi^2 - 17\psi + 11} = \frac{688 - 84\psi + 482\psi^2}{2075} \simeq 0.3839$$

is a normalisation constant.

The value of W_2 equals $-\log\lambda_{\max}$, where $\lambda_{\max}$ is the unique real root of the polynomial

$$t^9 + (3\psi^2 - 2\psi)t^7 + (2 - 3\psi - \psi^2)t^6 + (4\psi - 4\psi^2 - 1)t^5$$
$$+(7 - 14\psi + 3\psi^2)t^3 + (15 - 40\psi + 24\psi^2)t^2$$
$$+(24 - 33\psi - 16\psi^2)t + (-16 + 56\psi - 49\psi^2)\,.$$

This gives $W_2 \simeq 0.54922$. The algorithm in this example is the best among algorithms with $|\mathcal{U}| \le 3$ according to criteria ϱ and R, and in particular is significantly better than the GS algorithm.

Example 5.9 *This example is defined by Table 8.6. We have*

$$\phi_x^{(1)}(x) = \frac{36}{59}\mathrm{I}_{[\frac{1}{8},\frac{1}{4}]}(x) + \frac{28}{59}\mathrm{I}_{[\frac{1}{4},\frac{1}{2}]}(x) + \frac{16}{59}\mathrm{I}_{[\frac{1}{2},\frac{5}{8}]}(x)\,,$$
$$\phi_x^{(2)}(x) = \frac{36}{59}\mathrm{I}_{[\frac{1}{6},\frac{1}{3}]}(x) + \frac{30}{59}\mathrm{I}_{[\frac{1}{3},\frac{1}{2}]}(x) + \frac{18}{59}\mathrm{I}_{[\frac{1}{2},\frac{2}{3}]}(x)\,,$$
$$\phi_x^{(3)}(x) = \frac{24}{59}\mathrm{I}_{[\frac{1}{4},\frac{3}{4}]}(x)\,,$$
$$\phi_x^{(4)}(x) = \frac{18}{59}\mathrm{I}_{[\frac{1}{3},\frac{1}{2}]}(x) + \frac{30}{59}\mathrm{I}_{[\frac{1}{2},\frac{2}{3}]}(x) + \frac{36}{59}\mathrm{I}_{[\frac{2}{3},\frac{5}{6}]}(x)\,,$$
$$\phi_x^{(5)}(x) = \frac{12}{59}\mathrm{I}_{[\frac{1}{2},\frac{3}{4}]}(x) + \frac{20}{59}\mathrm{I}_{[\frac{3}{4},\frac{7}{8}]}(x)\,.$$

$$R = \frac{7}{12} \simeq 0.5833$$
$$\varrho = \frac{28}{59}\log 2 + \frac{12}{59}\log 3 \simeq 0.5524$$
$$W_2 = -\log\lambda_{\max} \simeq 0.5452\,,$$

where $\lambda_{\max}$ is the maximal real root of the polynomial

$$20736t^{10} - 5904t^8 - 1296t^7 + 400t^6 + 108t^5 - 36t^4 - 12t^3 - 4t^2 - t + 1\,.$$

These values are again better than for the GS algorithm.

Example 5.10 *Consider the Example of Table 8.9. The process can be initialised with* $(v^{(\alpha)}, v'^{(\alpha)}) = (0,1)$. *Consider the pairs* $(v^{(\beta)}, v'^{(\beta)}) = (1, 1/2)$, $(v^{(\gamma)}, v'^{(\gamma)}) = (0, 1/2)$. *We then use Table 5.8 to initialise Table 8.9. We can easily check that this initialisation has no influence on the asymptotic characteristics of the algorithm.*

v	v'	$L(v,v')$	$R(v,v')$
$v^{(\alpha)}$	$v'^{(\alpha)}$	$v^{(\beta)}$	$v^{(\gamma)}$
$v^{(\beta)}$	$v'^{(\beta)}$	$v^{(\beta)}$	$v_{(0,1)}$
$v^{(\gamma)}$	$v'^{(\gamma)}$	$v_{(0,0)}$	$v^{(\gamma)}$

Table 5.8 *Initialisation of Table 8.9*

In this example, the asymptotic log-rate ϱ *tends to the optimal one* ($\varrho = \log 2 \simeq 0.693147$) *when the number of states is increased to infinity. We take* $u'^{(k)} = u^{(k)}/2$ *when* $u^{(k)} = 1/(1+2^k)$ *and* $u'^{(k)} = (1+u^{(k)})/2$ *when* $u^{(k)} = 1 - 1/(1+2^k)$.

An important feature of this example is that the invariant measures $\phi_x^{(k,s)}$ *conditioned on any given state* (k,s) *are uniform between* $u^{(k,s)}$ *and* $u'^{(k,s)}$. *The ergodic probabilities of the events* (L) *and* (R) *are then equal to 1/2 for* $k < K$. *When* $k = K$, *the ergodic probability of the transition* $(K,s) \longrightarrow (K-1,s)$ *is one. The asymptotic rates can then be calculated analytically,*

$$
\begin{aligned}
R(K) &= \frac{1}{1+2^K}\left(2^{K-1} + \frac{1}{2+2^K} + \frac{1+2^{K-1}}{1+2^K}\right) \\
&\quad + \frac{2^{K-3}}{1+2^K}\sum_{k=0}^{K-2}\frac{1}{2^k(1+2^k)},
\end{aligned}
$$

and

$$
\varrho(K) = \frac{2^K}{1+2^K}\log 2,
$$

which tends to $\log 2$ *as* K *increases. The value of* $W_2 = -\log\lambda_{\max}$ *is obtained from* $\lambda_{\max} = t^*/4$, *where* t^* *is the largest real root of the polynomial*

$$
t^{K+2} - 2t^{K+1} - 4t^K + 8t^{K-1} + t - 4 .
$$

It satisfies $\lambda_{\max} > 1/2$ *and tends to* $1/2$ *as* K *increases.*

We shall compute now the asymptotic values of the performance characteristics for the GS4 and GS4_0 algorithms. In particular, we shall see that W_1 is not sensitive to the choice of ϵ in the GS4 algorithm (and is thus the same for GS4 and GS4_0), whereas W_2 and W_∞ are. This reinforces our view that $\mathbf{E} \log \mathbf{L}_n$ is not a suitable criterion for evaluating the performance of an algorithm.

Lyapunov exponent and ergodic log-rate

The Lyapunov exponent of the dynamical system (5.37) is defined by

$$\Lambda = \lim_{n\to\infty} \frac{1}{n} \sum_{k=1}^{n} \log |T'(z_k)| , \tag{5.39}$$

if this limit exists and is the same for almost all z_1. Birkhoff's ergodic Theorem, see, *e.g.*, Keane (1991), implies that Λ exists and is given by

$$\Lambda = - \sum_{i=1}^{12} \bar{\pi}_i \log R_i \simeq 0.63006 , \tag{5.40}$$

where $\bar{\pi}_i$ and R_i respectively correspond to the invariant probability and reduction rate for state S_i, given by Table 5.7 and (5.38). This expression for Λ also follows from Theorems 4.1, 4.2 and 4.7, which give $W_1 = \varrho = \Lambda \simeq 0.63006$.

Asymptotic behaviour of the expected length

As we shall see below, the value of W_γ, $\gamma \neq 1$, given by (4.9), depends on the choice of ϵ whereas it was not the case for W_1. For that reason, we consider the GS4 and GS4_0 algorithms separately.

GS4 algorithm: $\epsilon = (1-a)/2$ The initial probabilities $\pi_1^{(1)}, \pi_7^{(1)}$, $\pi_8^{(1)}$ and $\pi_{12}^{(1)}$ are equal to zero, so that, from the structure of the Markov chain,

$$\pi_1^{(k)} = \pi_7^{(k)} = \pi_8^{(k)} = \pi_{12}^{(k)} = 0 , \ \ \forall k \geq 1 .$$

We can thus consider a reduced Markov chain with eight states,

$$S'_1 = S_2 , \ S'_2 = S_3 , \ S'_3 = S_4 , \ S'_4 = S_5 , \ S'_5 = S_6 , \\ S'_6 = S_9 , \ S'_7 = S_{10} , \ S'_8 = S_{11} , \tag{5.41}$$

and the associated eight intervals $\mathcal{I}'_i$. Its transition probability matrix $\boldsymbol{\Pi}'$ is given by the corresponding submatrix of $\boldsymbol{\Pi}$. We have the following corollary of Theorem 4.6.

Corollary 5.4 *Assume that x^* has a uniform prior distribution on $[A, B]$. Then for any $\gamma \geq 0$, $\gamma \neq 1$, the algorithm GS4 is such that*

$$W_\gamma = \frac{1}{1-\gamma} \log \lambda_{\max}(\boldsymbol{\Pi}'_{\gamma-1}), \tag{5.42}$$

where $\lambda_{\max}(\boldsymbol{M})$ denotes the maximal eigenvalue of the matrix $\boldsymbol{M}$ and where

$$[\boldsymbol{\Pi}'_{\gamma-1}]_{ij} = [\boldsymbol{\Pi}'_{ij}]^\gamma . \tag{5.43}$$

In particular for $\gamma = 2$ and $\gamma = 1$ elementary but tedious calculations respectively give

$$W_2 = -\log \mu_2 \simeq 0.61273 ,$$

where $\mu_2 \simeq 0.54187$ is the largest positive root of the equation

$$\begin{aligned} 4t^6 - 8a^2t^4 + (-24a^3 + 54a^2 - 42a + 6)t^3 \\ +(-12a^3 + 18a^2 - 14a + 2)t^2 + (52a^3 - 102a^2 + 90a - 14)t \\ +68a^3 - 125a^2 + 99a - 15 = 0 , \end{aligned}$$

and

$$\varrho_0 = W_0 = \log \mu_0 \simeq 0.65103 , \tag{5.44}$$

where $\mu_0 \simeq 1.9175$ is the maximal root of $t^5 - t^4 - t^3 - 2t^2 + 2 = 0$.

GS4$_0$ algorithm: $\epsilon = 0$ In that case we can only obtain a bound on $\liminf_{N\to\infty} \frac{1}{N} \log \mathbf{E} L_n^\gamma$.

Theorem 5.6 *Assume that x^* has a uniform distribution on $[A, B]$. Then, for any γ and $n > 3$ the GS4$_0$ algorithm is such that*

$$\frac{\log \mathbf{E} L_n^\gamma}{n-2} \geq (1+\gamma)\log(1-a) + \frac{\gamma \log L_0}{n-2} + \frac{1}{n-2} \log \frac{(1-c)^{1+\gamma}}{(1-a)^2} , \tag{5.45}$$

and

$$\liminf_{n\to\infty} \frac{1}{n} \log \mathbf{E} L_n^\gamma \geq \max\left[(1+\gamma)\log(1-a) ,\ \log \lambda_{\max}(\boldsymbol{\Pi}'_\gamma)\right] , \tag{5.46}$$

with a and c given by (5.34) and $\boldsymbol{\Pi}'_\gamma$ given by (5.43).

Proof. From Theorem 4.5, we have

$$\log \mathbf{E} L_n^\gamma = \gamma \log L_0 + \log \boldsymbol{p}_\gamma^T \boldsymbol{\Pi}_\gamma^{n-2} \boldsymbol{q}_\gamma ,$$

with $\boldsymbol{p}_\gamma, \boldsymbol{q}_\gamma$ and $\boldsymbol{\Pi}_\gamma$ given by (4.19). When $\epsilon = 0$, $[\boldsymbol{p}_\gamma]_1 = [\boldsymbol{p}_\gamma]_{12} = 0$, and due to the structure of the matrix $\boldsymbol{\Pi}$ we have

$$\log \mathbf{E} L_n^\gamma = \gamma \log L_0 + \log(\boldsymbol{p}_\gamma'')^T (\boldsymbol{\Pi}_\gamma'')^{n-2} \boldsymbol{q}_\gamma'' ,$$

where

$$\boldsymbol{p}_\gamma'' = \left([\boldsymbol{p}_\gamma]_2, [\boldsymbol{p}_\gamma]_3, \ldots, [\boldsymbol{p}_\gamma]_{11}\right)^T , \quad \boldsymbol{q}_\gamma'' = \left([\boldsymbol{q}_\gamma]_2, [\boldsymbol{q}_\gamma]_3, \ldots, [\boldsymbol{q}_\gamma]_{11}\right)^T ,$$

and $\boldsymbol{\Pi}_\gamma''$ is the submatrix of $\boldsymbol{\Pi}_\gamma$ corresponding to the states $S_2, \ldots, S_{11}$. We can reorder the states and rearrange S_7 and S_8 to the last positions, so as to obtain the following representation for $\boldsymbol{\Pi}_\gamma''$:

$$\boldsymbol{\Pi}_\gamma'' = \begin{pmatrix} \boldsymbol{\Pi}_\gamma' & 0 \\ \boldsymbol{G} & \boldsymbol{F} \end{pmatrix} ,$$

with

$$\begin{aligned} \boldsymbol{F} &= \begin{pmatrix} (1-a)^{1+\gamma} & 0 \\ 1 & 0 \end{pmatrix} , \\ \boldsymbol{G} &= \begin{pmatrix} 0 & 0 & 0 & 0 & a^{1+\gamma} & 0 & 0 & 0 \\ 0 & 0 & 0 & 0 & 0 & 0 & 0 & 0 \end{pmatrix} , \end{aligned}$$

which corresponds to transition probabilities (to the power $1+\gamma$) from states S_7 and S_8. Then $(\boldsymbol{\Pi}_\gamma'')^{n-2}$ takes the form

$$(\boldsymbol{\Pi}_\gamma'')^{n-2} = \begin{pmatrix} (\boldsymbol{\Pi}_\gamma')^{n-2} & 0 \\ \boldsymbol{G}_{n-2} & \boldsymbol{F}^{n-2} \end{pmatrix} ,$$

where $\boldsymbol{G}_{n-2}$ is a matrix with non-negative elements. With the same rearrangement of the states we define

$$\begin{aligned} [\boldsymbol{p}_\gamma]_{7,8} &= ([\boldsymbol{p}_\gamma]_7, [\boldsymbol{p}_\gamma]_8)^T = (0, \frac{(1-c)^{1+\gamma}}{4^\gamma})^T , \\ [\boldsymbol{q}_\gamma]_{7,8} &= ([\boldsymbol{q}_\gamma]_7, [\boldsymbol{q}_\gamma]_8)^T \\ &= (4^\gamma(1-a)^{\gamma-1}, 4^\gamma(2-5a+6a^2-2a^3)^\gamma)^T , \\ [\boldsymbol{r}_\gamma]_{7,8} &= ([\boldsymbol{p}_\gamma]_2, \ldots, [\boldsymbol{p}_\gamma]_6, [\boldsymbol{p}_\gamma]_9, \ldots, [\boldsymbol{p}_\gamma]_{11})^T , \\ [\boldsymbol{s}_\gamma]_{7,8} &= ([\boldsymbol{q}_\gamma]_2, \ldots, [\boldsymbol{q}_\gamma]_6, [\boldsymbol{q}_\gamma]_9, \ldots, [\boldsymbol{q}_\gamma]_{11})^T . \end{aligned}$$

Then, since all components involved are non-negative,

$$\begin{aligned} \log \mathbf{E} L_n^\gamma &\geq \gamma \log L_0 \\ &\quad + \log\left({[\boldsymbol{p}_\gamma]_{7,8}}^T \boldsymbol{F}^{n-2} [\boldsymbol{q}_\gamma]_{7,8} + {[\boldsymbol{r}_\gamma]_{7,8}}^T {\boldsymbol{\Pi}_\gamma'}^{n-2} [\boldsymbol{s}_\gamma]_{7,8} \right) , \end{aligned}$$

and thus

$$\log \mathbf{E} L_n^\gamma \geq \gamma \log L_0$$

$$+\max\left\{\log([\boldsymbol{p}_\gamma]_{7,8}{}^T\boldsymbol{F}^{n-2}[\boldsymbol{q}_\gamma]_{7,8}),\ \log([\boldsymbol{r}_\gamma]_{7,8}{}^T\boldsymbol{\Pi}'^{\,n-2}_\gamma[\boldsymbol{s}_\gamma]_{7,8})\right\}.$$

The matrix $\boldsymbol{F}$ can be decomposed into

$$\boldsymbol{F} = \begin{pmatrix} 1 & (1-a)^{1+\gamma} \\ 1 & 0 \end{pmatrix}^{-1} \begin{pmatrix} 0 & 0 \\ 0 & (1-a)^{1+\gamma} \end{pmatrix} \times \begin{pmatrix} 1 & (1-a)^{1+\gamma} \\ 1 & 0 \end{pmatrix},$$

which gives after elementary calculations

$$\log\left([\boldsymbol{p}_\gamma]_{7,8}^T\boldsymbol{F}^{n-2}[\boldsymbol{q}_\gamma]_{7,8}\right) = (1+\gamma)\log(1-a)^{n-2} + \log\frac{(1-c)^{1+\gamma}}{(1-a)^2},$$

and thus (5.45) and (5.46). ■

When $\gamma = 1$ we thus have the following:

Corollary 5.5 *For the GS4$_0$ algorithm one has when x^* has a prior distribution uniform on $[A, B]$*

$$\liminf_{n\to\infty}\frac{1}{n}\log \mathrm{E}L_n \geq 2\log(1-a) \simeq -0.43163.$$

Note that the bound $2\log(1-a)$ of Corollary 5.5 implies that the performances in terms of expected length of the uncertainty interval are significantly worse for the GS4$_0$ algorithm than for the GS algorithm, for which $\log \mathrm{E}L_n/n = [(n-1)\log\varphi]/n$, with $\log\varphi \simeq -0.48121$. Also note that (5.44) gives a lower bound for the topological entropy of the dynamical system associated with the GS4$_0$ algorithm, since the partition generated by GS4$_0$ is finer than that generated by GS4.

The zeta functions (4.25) for the GS4 and GS4$_0$ algorithms are respectively

$$\zeta(t) = \frac{1}{1-2t^2-3t^3-2t^4+2t^5+2t^6},$$

and

$$\zeta(t) = \frac{1}{1-t-2t^2-t^3+t^4+4t^5-2t^7}.$$

This gives the following values for the number of periodic trajectories of period n, see (4.24), for $n = 1, \ldots, 10$:

GS4: $0, 4, 9, 16, 20, 55, 98, 176, 351, 684,$

GS4$_0$: $1, 5, 10, 17, 21, 56, 99, 177, 352, 685.$

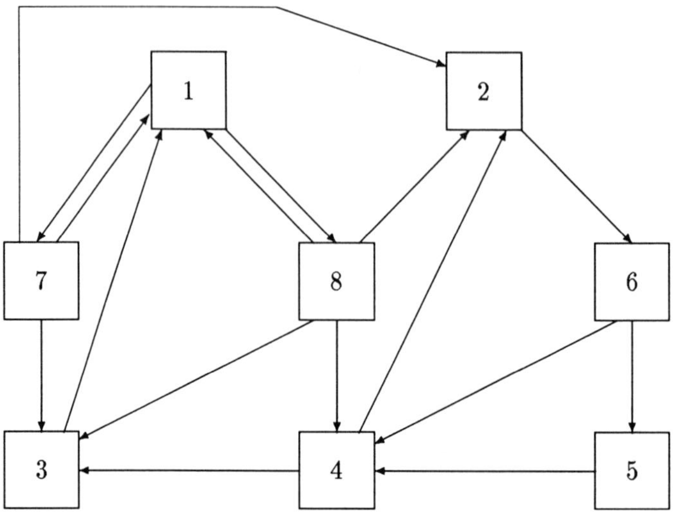

Figure 5.13 *Graph of transitions for states S'_j for the GS4 algorithm*

Finite-sample worst-case performance characteristics

The expression for $\log \mathtt{M}L_n$ can be computed from the determination of the worst cases. Again the two cases $\epsilon = (1-a)/2$ and $\epsilon = 0$ have to be considered separately.

Case a: $\epsilon = (1-a)/2$ We define the adjacency matrix for the states $S'_i, i = 1, \ldots, 8$ as

$$[\boldsymbol{\Pi}'_{-1}]_{ij} = \begin{cases} 1 & \text{if } [\boldsymbol{\Pi}']_{ij} > 0\,, \\ 0 & \text{otherwise}\,; \end{cases} \tag{5.47}$$

that is, $[\boldsymbol{\Pi}'_{-1}]_{ij} = 1$ if and only if state S'_j is reachable in one step from state S'_i. The corresponding transition graph is presented on Figure 5.13.

Theorem 5.7 *For the GS4 algorithm one has when $f(\cdot)$ is symmetric with respect to x^**

$$\forall n \geq 3\,, \quad \log \mathtt{M}L_n = (2m+1)\log(1-a) + m\log a' + m \log c + l_k\,,$$

where $k = (n-3)[mod\ 4]$, $m = (n-3-k)/4$ *and*

$$l_k = \begin{cases} 0 & \textit{if } k = 0 \\ \log(1-a) & \textit{if } k = 1 \\ 2\log(1-a) & \textit{if } k = 2 \\ 2\log(1-a) + \log a' & \textit{if } k = 3 \end{cases}$$

Proof. One can easily check that there are just two cycles which give the same asymptotic worst rate, namely:

$$S'_4 \longrightarrow S'_3 \longrightarrow S'_1 \longrightarrow S'_8 \longrightarrow S'_4 \longrightarrow \cdots$$
$$S'_4 \longrightarrow S'_2 \longrightarrow S'_6 \longrightarrow S'_5 \longrightarrow S'_4 \longrightarrow \cdots$$

with rate $a'c(1-a)^2$ for 4 iterations.

The initial distribution is concentrated on $\mathcal{I}'_6$ and $\mathcal{I}'_7$, *i.e.*, the initial state is S'_6 or S'_7. Starting at S'_6, where the rate is c, we go in one iteration to S'_4 or S'_5, where the rate is $1-a$. In the next iteration we then respectively enter one of the worst cycles described above, or go first from S'_5 to S'_4 and then enter one of the worst cycles. The later case gives the worst case.

Starting at S'_7, where the rate is c, we first go either to S'_1 or S'_2, which belong to the cycles above and have rate a'.

The worst path thus starts at S'_6:

$$S'_6 \longrightarrow S'_5 \longrightarrow S'_4 \longrightarrow \cdots \longrightarrow S'_4 \longrightarrow S'_3 \longrightarrow S'_1 \longrightarrow S'_8\,,$$

and stops at S'_4, S'_3, S'_7 or S'_8 depending on the value of N. Taking into account that the rate r_1 corresponding to the initial expansion of the interval is $1/c$, we have the stated result. ■

Corollary 5.6 *For the GS4 algorithm for functions symmetric with respect to* x^*, *one has*

$$\forall n \geq 38\,, \quad \forall x^* \in [A,B]\,, \quad L_n(x^*) < L_0\varphi^{n-1}$$
$$\forall n \geq 42\,, \quad \forall x^* \in [A,B]\,, \quad L_n(x^*) < \tfrac{L_0}{F_{n+1}}\,;$$

that is, the GS4 algorithm performs better than the GS (respectively Fibonacci) algorithm for any $n \geq 38$ *(respectively 42) and any* x^* *in* $[A,B]$.

There are only two values for x^* that make the algorithm stay in the worst-case path forever. They can be determined as follows. One can check that the point

$$\hat{z} = -\frac{2}{31}a^3 - \frac{3}{31}a^2 + \frac{19}{31}a + \frac{3}{31} \simeq 0.21163$$

is the only point in $\mathcal{I}'_4$ such that $T^4(\hat{z}) = \hat{z}$ and $T(\hat{z}) \in \mathcal{I}'_3$, $T^2(\hat{z}) \in \mathcal{I}'_1$ and $T^3(\hat{z}) \in \mathcal{I}'_8$. With $\hat{z}$ is associated a unique point z_1 in $\mathcal{I}'_6$ such that $T(z_1) \in \mathcal{I}'_5$ and $T^2(z_1) = \hat{z}$. It is given by

$$z_1 = \frac{7}{31}a^3 - \frac{5}{31}a^2 - \frac{9}{62}a + \frac{41}{62} \simeq 0.62869\,.$$

To z_1 correspond two possible values for x^*, given by $x^* = x_1 L_1 + A_1$, with $L_1 = (2-a)(B-A)$, $A_1 = A - \frac{1-a}{2}(B-A)$ and $x_1 = 2z_1 - 1$ or $x_1 = 2 - 2z_1$; that is,

$$x^* \simeq A + 0.061843(B-A)\,, \text{ or } x^* \simeq A + 0.938157(B-A)\,.$$

GS4$_0$ algorithm: $\epsilon = 0$ When $\epsilon = 0$, we need to consider all states $S_i, i = 1, \ldots, 12$. The worst cycle, corresponding to $x^* = A$ or $x^* = B$, is then

$$S_7 \longrightarrow S_7 \cdots$$

with rate $1 - a$ for 1 iteration. We then obtain the following property.

Theorem 5.8 *For the GS4$_0$ algorithm for functions symmetric with respect to x^*, one has*

$$\forall n \geq 2\,, \quad \log \mathtt{M}\mathcal{L}_n = \log c + (n-2)\log(1-a)\,.$$

Proof. The algorithm is initialised in S_8 or S_9 or S_{10}. The worst path is

$$S_8 \longrightarrow S_7 \longrightarrow \cdots \longrightarrow S_7$$

which gives the result. ∎

Asymptotic behaviour of worst-case performances

We consider now the asymptotic worst-case characteristic W_∞. Again, the two cases $\epsilon = (1-a)/2$ and $\epsilon = 0$ must be treated separately.

GS4 algorithm: $\epsilon = (1-a)/2$ From Theorem 5.7, we obtain the asymptotic expression

$$\lim_{n\to\infty} \frac{1}{n} \log \mathtt{ML}_n = \frac{1}{4}\log(a(1-a)^2) \simeq -0.51773\,.$$

A crucial point here is that this value is less than $\log \varphi \simeq -0.48121$, which corresponds to the performance of the GS algorithm.

GS4$_0$ algorithm: $\epsilon = 0$ Theorem 5.8 now gives the asymptotic expression

$$\lim_{n\to\infty} \frac{1}{n} \log \mathtt{ML}_n = \log(1-a) \simeq -0.21582\,.$$

This value is now much larger than $\log \varphi$. However, we still have convergence to 1 for P_N^{GS} and P_N^F respectively defined by (4.11, 4.12); see Theorem 4.3.

Summary

The asymptotic performances of the GS, GS4 and $GS4_0$ algorithms are summarized in Table 5.9 and Figure 5.14.

	GS	$GS4_0$	GS4	upper bound
W_0	0.48121	0.65103	0.65103	$\log 2$
W_1	0.48121	0.63006	0.63006	$\log 2$
W_2	0.48121	0.43163	0.61273	$\log 2$
W_∞	0.48121	0.21582	0.51773	$\log 2$

Table 5.9 *Asymptotic performance characteristics of the GS, $GS4_0$ and GS4 algorithms ($\log 2 \simeq 0.6930$)*

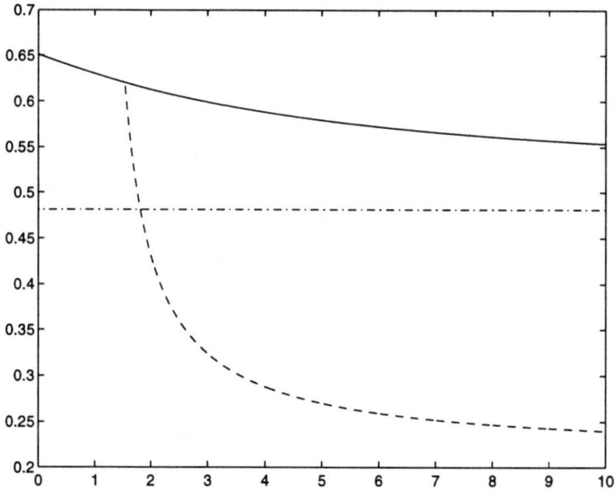

Figure 5.14 *Evolution of W_γ as a function of γ for the GS4 (full line), $GS4_0$ (dashed line) and GS (dash-dotted line) algorithms. The dashed and full lines are confounded for $\gamma < 1.53$*

Note that for $\gamma < \gamma^* \simeq 1.5335$, W_γ does not indicate any difference between the behaviours of the GS4 and $GS4_0$ algorithms. This reinforces our opinion that W_1 (and therefore Kolmogorov entropy) is not always sensitive enough to detect differences in asymptotic behaviours.

The performances in terms of $\lim_{n\to\infty} \frac{1}{n} \log \bar{L}_n^{1-\alpha}$ are not indicated in Table 5.9 since this characteristic coincides with $\lim_{n\to\infty} \frac{1}{n} \mathrm{E} \log L_n$; see Theorem 4.3. The fact that $-\log 2$ is a lower bound for $\lim_{n\to\infty} \frac{1}{n} \mathrm{E} \log L_n$ is proved in Section 5.4.1, and an algorithm that achieves this lower bound for locally symmetric functions satisfying (5.48) is detailed in Wynn and Zhigljavsky (1995). The fact that $-\log 2$ is a lower bound for $\lim_{n\to\infty} \frac{1}{n} \log \mathrm{E} L_n$ follows from Jensen's inequality:

$$\log \mathrm{E}_{x^*} \{L_n(x^*)\} \geq \mathrm{E}_{x^*} \{\log L_n(x^*)\}.$$

This also gives an upper bound for the topological entropy, since one cannot divide the interval by more than 2 at each iteration.

A family of algorithms with performances $\lim_{n\to\infty} \frac{1}{n} \log \mathrm{E} L_n$ arbitrarily close to the bound $-\log 2$ for functions symmetric with respect to x^* is presented in Example 5.10. However, these algorithms are mainly of theoretical interest since their finite sample behaviour is inferior. The lower bound $-\log 2$ for $\lim_{n\to\infty} \frac{1}{n} \log \mathrm{M} L_n$ simply follows from

$$\max_{x^* \in [A,B]} \{\log L_N(x^*)\} \geq \mathrm{E}_{x^*} \{\log L_N(x^*)\}.$$

The existence of second-order algorithms achieving this bound for functions symmetric with respect to x^* remains an open question.

Table 5.9 shows that the GS4 algorithm has much better asymptotic performances than the GS algorithm. This alone is not enough, however, to give GS4 some practical interest. One should also consider performances obtained for finite n, for symmetric and asymmetric functions. This is done in the next section.

5.8 Comparisons of GS, GS4, $GS4_0$ and window algorithms

5.8.1 *Finite sample performance characteristics*

We first assume that $f(\cdot)$ is symmetric with respect to x^*. Robustness with respect to asymmetry will be investigated in Section 5.8.2.

As shown above, the criteria $\mathrm{E} \log L_n$, $\log \mathrm{E} L_n^\gamma$ and $\log \mathrm{M} L_n$ can be calculated analytically for the GS, GS4 and $GS4_0$ algorithms. The criteria $\bar{L}_n^{1-\alpha}$, see (4.10), P_n^{GS} and P_n^F, see (4.11) and (4.12), are difficult to compute analytically, but can be evaluated with any arbitrary precision for any reasonable n. We simply need to compute the value of L_n and the probability $\pi_{i_1}^{(1)}\pi_{i_1 i_2}\cdots\pi_{i_{n-2}i_{n-1}}$ associated with any sequence of states $S_{i_1}, S_{i_2}, \ldots, S_{i_{n-1}}$. Note that from Corollary 5.6, $P_n^{GS} = 1, n \geq 38$ and $P_n^F = 1, n \geq 42$ for the GS4 algorithm.

Table 5.10 presents the numerical performances of the GS4 algorithm ($\epsilon = (1-a)/2$) with respect to all criteria above for $1 \leq n \leq 30$. Note the expansion of the initial interval, that is, $L_1 > L_0$.

n	φ^{n-1}	$\frac{1}{F_{n+1}}$	EL_n	ML_n	P_n^{GS}	P_n^F	$\bar{L}_n^{0.99}$
1	1.0000	1.0000	1.8059	1.8059	.0000	.0000	1.8059
2	.6180	.5000	1.0000	1.0000	.0000	.0000	1.0000
3	.3820	.3333	.6463	.8059	.3506	.0000	.8059
4	.2361	.2000	.3862	.6494	.3506	.3506	.6494
5	.1459	.1250	.2117	.5234	.1361	.1361	.5234
6	9.017 .$^{-2}$	7.692 .$^{-2}$	.1064	.1835	.3555	[illegible]	.1835
7	5.573 .$^{-2}$	4.762 .$^{-2}$	6.076 .$^{-2}$	.1016	.6100	.6100	.1016
8	3.444 .$^{-2}$	2.941 .$^{-2}$	3.345 .$^{-2}$	8.188 .$^{-2}$	.6100	.6100	8.188 .$^{-2}$
9	2.129 .$^{-2}$	1.818 .$^{-2}$	1.764 .$^{-2}$	6.598 .$^{-2}$	.8421	.3823	6.598 .$^{-2}$
10	1.316 .$^{-2}$	1.124 .$^{-2}$	9.511 .$^{-3}$	2.313 .$^{-2}$	.6196	.6196	2.313 .$^{-2}$
11	8.131 .$^{-3}$	6.944 .$^{-3}$	5.266 .$^{-3}$	1.281 .$^{-2}$	.8207	.8207	1.281 .$^{-2}$
12	5.025 .$^{-3}$	4.292 .$^{-3}$	2.835 .$^{-3}$	1.032 .$^{-2}$	.9437	.8207	1.032 .$^{-2}$
13	3.106 .$^{-3}$	2.653 .$^{-3}$	1.521 .$^{-3}$	8.318 .$^{-3}$	.9437	.9437	3.618 .$^{-3}$
14	1.919 .$^{-3}$	1.639 .$^{-3}$	8.300 .$^{-4}$	2.916 .$^{-3}$	.8080	.8080	2.004 .$^{-3}$
15	1.186 .$^{-3}$	1.013 .$^{-3}$	4.515 .$^{-4}$	1.615 .$^{-3}$	.9267	.9267	1.615 .$^{-3}$
16	7.331 .$^{-4}$	6.261 .$^{-4}$	2.433 .$^{-4}$	1.301 .$^{-3}$	.9817	.9817	1.301 .$^{-3}$
17	4.531 .$^{-4}$	3.870 .$^{-4}$	1.318 .$^{-4}$	1.049 .$^{-3}$	.9817	.9817	4.562 .$^{-4}$
18	2.800 .$^{-4}$	2.392 .$^{-4}$	7.166 .$^{-5}$	3.676 .$^{-4}$	.9978	.9134	2.526 .$^{-4}$
19	1.731 .$^{-4}$	1.478 .$^{-4}$	3.879 .$^{-5}$	2.036 .$^{-4}$	.9725	.9725	2.036 .$^{-4}$
20	1.070 .$^{-4}$	9.136 .$^{-5}$	2.098 .$^{-5}$	1.640 .$^{-4}$	.9943	.9943	7.136 .$^{-5}$
21	6.611 .$^{-5}$	5.646 .$^{-5}$	1.138 .$^{-5}$	1.322 .$^{-4}$	.9995	.9943	3.952 .$^{-5}$
22	4.086 .$^{-5}$	3.490 .$^{-5}$	6.173 .$^{-6}$	4.634 .$^{-5}$	.9994	.9994	3.184 .$^{-5}$
23	2.525 .$^{-5}$	2.157 .$^{-5}$	3.341 .$^{-6}$	2.566 .$^{-5}$	.9903	.9903	1.116 .$^{-5}$
24	1.561 .$^{-5}$	1.333 .$^{-5}$	1.811 .$^{-6}$	2.068 .$^{-5}$	.9983	.9983	6.182 .$^{-6}$
25	9.645 .$^{-6}$	8.238 .$^{-6}$	9.817 .$^{-7}$	1.667 .$^{-5}$	.9999	.9999	4.982 .$^{-6}$
26	5.961 .$^{-6}$	5.091 .$^{-6}$	5.318 .$^{-7}$	5.843 .$^{-6}$	1.000	.9999	4.015 .$^{-6}$
27	3.684 .$^{-6}$	3.147 .$^{-6}$	2.881 .$^{-7}$	3.235 .$^{-6}$	1.000	.9967	1.407 .$^{-6}$
28	2.277 .$^{-6}$	1.945 .$^{-6}$	1.561 .$^{-7}$	2.607 .$^{-6}$	.9995	.9995	7.793 .$^{-7}$
29	1.407 .$^{-6}$	1.202 .$^{-6}$	8.462 .$^{-8}$	2.101 .$^{-6}$	1.000	1.000	6.280 .$^{-7}$
30	8.697 .$^{-7}$	7.428 .$^{-7}$	4.584 .$^{-8}$	7.366 .$^{-7}$	1.000	1.000	2.202 .$^{-7}$

Table 5.10 *Performances of the GS4 algorithm with* $L_0 = 1$ *(*$.^j$ *means* 10^j*)*

Figure 5.15 gives the evolution of some performance characteristics as function of n for the GS4 algorithm.

Table 5.11 presents the performances achieved for the $GS4_0$ algo-

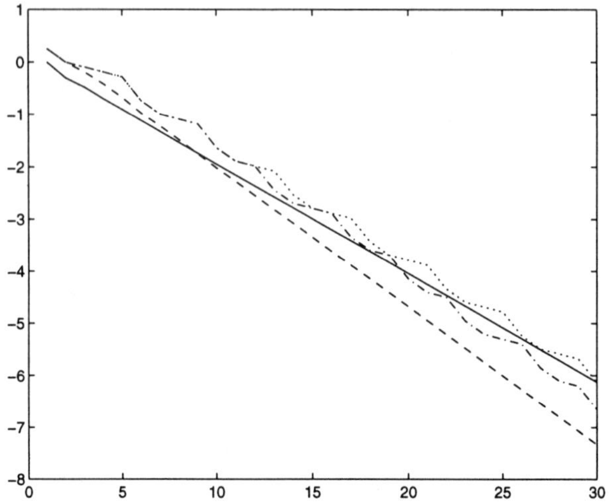

Figure 5.15 *Decimal logarithm of various performance characteristics for the GS4 algorithm as a function of n: full line for* $1/F_{n+1}$, *dashed line for* $\mathtt{E}L_n$, *dash-dotted line for* $\bar{L}_n^{0.99}$ *and dotted line for* $\mathtt{M}L_n$

rithm ($\epsilon = 0$). The comparison with Table 5.10 stresses the importance of expanding the initial interval for the finite sample behaviour. The fact that this expansion is also important asymptotically has already been demonstrated in Section 5.7.2.

Consider now the window algorithm of Sections 2.2.3 and 5.6.2, defined by (2.13, 2.14), with $\epsilon = 0.3772$ and $w = 0.15$.

When $f(\cdot)$ is symmetric with respect to x^*, the decision about the right or left deletion in (3.11) only depends on the location of x^* in the interval. For any fixed n, the initial interval $[A, B]$ can thus be partitioned into a union of disjoint sub-intervals $\cup_i \mathcal{I}_n^i$, such that the behaviour of the algorithm is the same up to iteration n for all x^* in $\mathcal{I}_n^i$. Note that the cardinality of this partition grows exponentially with n (it is already 11,760 for $n = 16$). Let $I_1^1 = [A, B], U_1^1 = U_1 = E_1$. At iteration 2, I_1^1 is split into two subintervals $\mathcal{I}_2^1 = [A, (A+B)/2), \mathcal{I}_2^2 = [(A+B)/2, B]$, corresponding respectively to right and left deletion. In the first case U_1^1 is updated into $U_2^1 = U_1^1 - w(E_1' - A_1)$, in the second into $U_2^2 = E_1'$. Using similar arguments, we can track the path of the end points of $\mathcal{I}_n^i$, together with U_n^i, in order to construct the partition $\cup_i \mathcal{I}_n^i$ for any n: at iteration n, each interval $\mathcal{I}_n^i$ is updated into either $\mathcal{I}_{n+1}^i$

n	EL_n	ML_n	P_n^{GS}	P_n^F	$\bar{L}_n^{0.99}$
1	1	1	0	0	1
2	.5537	.5537	0	0	.5537
3	.3973	.4463	.1941	.1941	.4463
4	.2789	.3596	.3506	.3506	.3596
5	.1943	.2898	.4766	.3506	.2898
6	.1336	.2336	.4766	.4766	.2336
7	$9.069\ 10^{-2}$	.1882	.3378	.3378	.1882
8	$6.098\ 10^{-2}$	.1517	.4663	.4663	.1517
9	$4.078\ 10^{-2}$	.1222	.5699	.5699	.1222
10	$2.712\ 10^{-2}$	$9.851\ 10^{-2}$	.6734	.5699	$9.851\ 10^{-2}$
15	$3.350\ 10^{-3}$	$3.348\ 10^{-2}$	.7209	.7209	$3.348\ 10^{-2}$
20	$3.972\ 10^{-4}$	$1.138\ 10^{-2}$	.8173	.8173	$1.138\ 10^{-2}$
25	$4.637\ 10^{-5}$	$3.869\ 10^{-3}$	.8830	.8205	$1.156\ 10^{-3}$
30	$5.381\ 10^{-6}$	$1.315\ 10^{-3}$	.8838	.8838	$5.110\ 10^{-5}$

Table 5.11 *Performances of the $GS4_0$ algorithm* ($L_0 = 1$)

or the union of two disjoint intervals, depending on the location of U_n^i with respect to $\mathcal{I}_n^i$. Once the partition is constructed, the values of all criteria above can easily be computed with any given arbitrary precision. The results are summarized in Table 5.12.

Figure 5.16 gives the evolution of some performance characteristics as functions of n for the window algorithm.

Table 5.13 presents the value of N required for the corresponding characteristic to reach the precision indicated. For instance, the Fibonacci algorithm requires 30 function evaluations to reduce the length of the initial interval by a factor 10^6, while the GS4 (window) algorithm requires respectively 25 (25) and 28 (27) evaluations to achieve the same precision, on the average and with probability 0.99.

The results above show that the GS4 and window algorithms have significantly better performances than the GS and Fibonacci algorithms for reasonable values of n. The performances of the window algorithm are slightly better than those of the GS4 algorithm. However, these results are obtained under the assumption that $f(\cdot)$ is symmetric with respect to x^*, and it is therefore important to check the validity of these conclusions for asymmetric functions.

n	φ^{n-1}	$\frac{1}{F_{n+1}}$	EL_n	ML_n	P_n^{GS}	P_n^F	$\bar{\mathrm{L}}_n^{0.99}$
3	.3820	.3333	.6084	.7456	.0000	.0000	.7456
4	.2361	.2000	.3593	.5943	.2135	.0000	.5943
5	.1459	.1250	.1960	.4825	.1958	.1958	.4825
6	9.017 .$^{-2}$	7.692 .$^{-2}$	9.760 .$^{-2}$	.1615	.4445	.4445	.1615
7	5.573 .$^{-2}$	4.762 .$^{-2}$	5.433 .$^{-2}$	9.660 .$^{-2}$	.6887	.5864	9.660 .$^{-2}$
8	3.444 .$^{-2}$	2.941 .$^{-2}$	2.939 .$^{-2}$	7.237 .$^{-2}$	.7993	.5570	7.237 .$^{-2}$
9	2.129 .$^{-2}$	1.818 .$^{-2}$	1.546 .$^{-2}$	5.788 .$^{-2}$	.7893	.7318	5.788 .$^{-2}$
10	1.316 .$^{-2}$	1.124 .$^{-2}$	8.326 .$^{-3}$	1.954 .$^{-2}$	.7581	.7211	1.937 .$^{-2}$
11	8.131 .$^{-3}$	6.944 .$^{-3}$	4.516 .$^{-3}$	1.495 .$^{-2}$	.8831	.8831	1.159 .$^{-2}$
12	5.025 .$^{-3}$	4.292 .$^{-3}$	2.420 .$^{-3}$	9.271 .$^{-3}$	.9619	.9376	8.657 .$^{-3}$
13	3.106 .$^{-3}$	2.653 .$^{-3}$	1.295 .$^{-3}$	6.944 .$^{-3}$	.9900	.9254	3.044 .$^{-3}$
14	1.919 .$^{-3}$	1.639 .$^{-3}$	6.983 .$^{-4}$	2.344 .$^{-3}$	.9806	.9535	2.209 .$^{-3}$
15	1.186 .$^{-3}$	1.013 .$^{-3}$	3.763 .$^{-4}$	1.807 .$^{-3}$	.9736	.9572	1.302 .$^{-3}$
16	7.331 .$^{-4}$	6.261 .$^{-4}$	2.020 .$^{-4}$	1.114 .$^{-3}$	.9890	.9890	8.506 .$^{-4}$
17	4.531 .$^{-4}$	3.870 .$^{-4}$	1.085 .$^{-4}$	8.432 .$^{-4}$	.9982	.9974	3.476 .$^{-4}$
18	2.800 .$^{-4}$	2.392 .$^{-4}$	5.840 .$^{-5}$	2.877 .$^{-4}$	.9990	.9938	2.084 .$^{-4}$
19	1.731 .$^{-4}$	1.478 .$^{-4}$	3.141 .$^{-5}$	2.168 .$^{-4}$	.9993	.9934	1.396 .$^{-4}$
20	1.070 .$^{-4}$	9.136 .$^{-5}$	1.688 .$^{-5}$	1.337 .$^{-4}$	.9977	.9966	5.590 .$^{-5}$
21	6.611 .$^{-5}$	5.646 .$^{-5}$	9.074 .$^{-6}$	1.012 .$^{-4}$	.9996	.9996	3.656 .$^{-5}$
22	4.086 .$^{-5}$	3.490 .$^{-5}$	4.881 .$^{-6}$	3.477 .$^{-5}$	1.000	1.000	2.140 .$^{-5}$
23	2.525 .$^{-5}$	2.157 .$^{-5}$	2.625 .$^{-6}$	2.633 .$^{-5}$	.9999	.9998	9.926 .$^{-6}$
24	1.561 .$^{-5}$	1.333 .$^{-5}$	1.411 .$^{-6}$	1.623 .$^{-5}$	1.000	.9994	5.928 .$^{-6}$
25	9.645 .$^{-6}$	8.238 .$^{-6}$	7.587 .$^{-7}$	1.228 .$^{-5}$	.9999	.9999	3.447 .$^{-6}$
26	5.961 .$^{-6}$	5.091 .$^{-6}$	4.080 .$^{-7}$	4.173 .$^{-6}$	1.000	1.000	1.646 .$^{-6}$
27	3.684 .$^{-6}$	3.147 .$^{-6}$	2.194 .$^{-7}$	3.158 .$^{-6}$	1.000	1.000	9.468 .$^{-7}$
28	2.277 .$^{-6}$	1.945 .$^{-6}$	1.180 .$^{-7}$	1.947 .$^{-6}$	1.000	1.000	5.598 .$^{-7}$
29	1.407 .$^{-6}$	1.202 .$^{-6}$	6.342 .$^{-8}$	1.474 .$^{-6}$	1.000	1.000	2.908 .$^{-7}$
30	8.697 .$^{-7}$	7.428 .$^{-7}$	3.410 .$^{-8}$	5.065 .$^{-7}$	1.000	1.000	1.558 .$^{-7}$

Table 5.12 *Performances of the window algorithm with* $L_0 = 1$ *(*$.^j$ *means* 10^j*)*

5.8.2 *Robustness with respect to asymmetry*

We consider now the case where $f(\cdot)$ is only locally symmetric around x^*; that is,

$$\begin{aligned} C|x-x^*|^\beta - D|x-x^*|^{\beta+\omega} &\le \\ |f(x)-f(x^*)| &\le C|x-x^*|^\beta + D|x-x^*|^{\beta+\omega}, \end{aligned} \tag{5.48}$$

with $\beta > 0, \omega > 0, C > 0, D \ge 0$. Since the algorithm only uses function comparisons, $f(\cdot)$ can be scaled down so that we can assume that $C = 1$.

Figure 5.17 presents the graph of the functions $f_{0.5}(x-x^*)$ and $f_2(x-x^*)$, with

$$f_D(z) = \begin{cases} \frac{4}{27D^2} & \text{if } z \le -\frac{2}{3D} \\ z^2 + Dz^3 & \text{if } -\frac{2}{3D} < z, \end{cases}$$

which can be considered as the worst uni-extremal function in the class above with respect to the symmetry condition, for a given

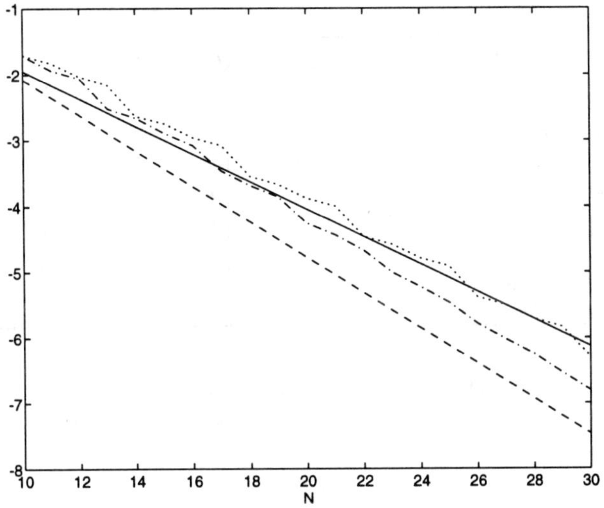

Figure 5.16 *Decimal logarithm of various performance characteristics for the window algorithm as a function of n: full line for* $1/F_{n+1}$*, dashed line for* $\mathrm{E}L_n$*, dash-dotted line for* $\bar{L}_n^{0.99}$ *and dotted line for* $\mathrm{M}L_n$

value D, when $\beta = 2, \omega = 1$. The functions $f_D(\cdot)$ are constant for $x \leq \bar{x} = -2/(3D)$; however, since we delete $[A_n, U_n]$ when $f(U_n) \geq f(V_n)$ this has no effect on the behaviour of the algorithm.

Numerical simulations with x^* uniformly distributed in $[A, B]$ were carried out, with the asymmetry parameter D varying from 0 to 10. Figure 5.18 presents the evolution of the empirical values of the performance characteristics EL_{30} and ML_{30} as functions of D for the GS4 algorithm. Note that EL_{30} remains much smaller than $1/F_{31}$, even for functions quite asymmetric with respect to x^*. Note in particular from Figure 5.17 that $f_D(\cdot)$ is already very asymmetric when $D = 2$. The value of ML_{30} jumps from $\mathrm{ML}_{30}^{(1)} \simeq 7.366\ 10^{-7}$ to $\mathrm{ML}_{30}^{(2)} = [(1-a)/a']\mathrm{ML}_{30}^{(1)} \simeq 1.693\ 10^{-6}$ at $D \simeq 1.35$. This jump corresponds to replacing one transition with rate a' by another one with rate $1 - a$ in the worst path through the graph presented in Figure 5.13.

Figure 5.19 presents the evolution of the empirical probability P_n^F, given by (4.12), as a function of D and n. We have already

	precision	10^{-1}	10^{-2}	10^{-3}	10^{-4}	10^{-5}	10^{-6}
GS	φ^{n-1}	6	11	16	21	25	30
Fib.	$\frac{1}{F_{n+1}}$	6	11	16	20	25	30
GS4	$\mathbf{EL}_n$	7	10	14	18	22	25
GS4	$\mathbf{ML}_n$	8	13	18	22	26	30
GS4	$\bar{\mathbf{L}}_n^{0.99}$	8	13	17	20	24	28
wind.	$\mathbf{EL}_n$	6	10	14	18	21	25
wind.	$\mathbf{ML}_n$	7	12	17	22	26	30
wind.	$\bar{\mathbf{L}}_n^{0.99}$	7	12	16	20	23	27

Table 5.13 *Number n of function evaluations required to achieve a given precision for the Fibonacci, GS, GS4 and window algorithms* ($L_0 = 1$)

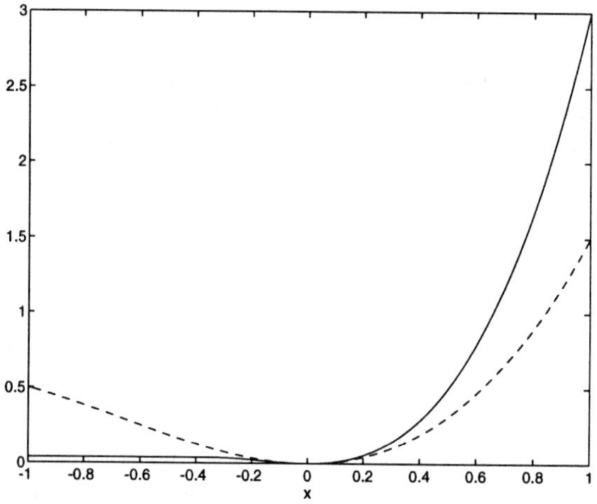

Figure 5.17 *Graphs of* $f_{0.5}(\cdot)$ *and* $f_2(\cdot)$

seen that $P_n^F = 1$ for $n \geq 42$ when $D = 0$, and this figure shows that the good performances of the GS4 algorithm remain fairly stable while asymmetry increases.

Figures 5.20 and 5.21 present the same information for the window algorithm as Figures 5.18 and 5.19 for the GS4 algorithm.

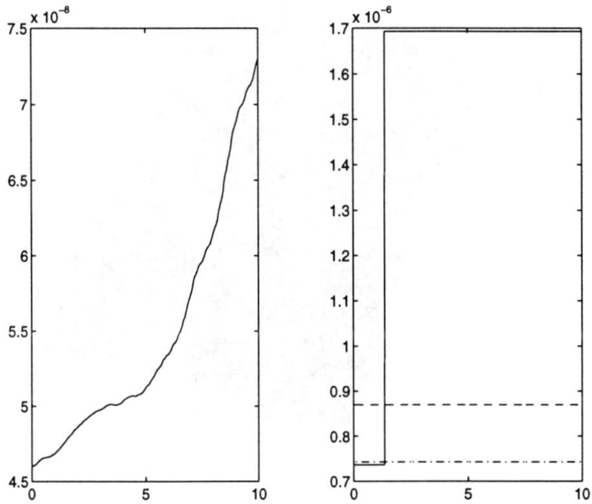

Figure 5.18 *Evolution of* $\mathrm{E}L_{30}$ *(left) and* $\mathrm{M}L_{30}$ *(right) as functions of* D *for the GS4 algorithm. The dashed and dash-dotted horizontal lines on the right respectively correspond to* φ^{29} *and* $1/F_{31}$*, that is, to the GS and Fibonacci algorithms*

One can conclude from all these results that the GS4 and window algorithms are interesting alternatives to the well known and widely used GS and Fibonacci algorithms. Indeed, their good performances for moderate values of n make them attractive for practical implementation in optimisation software.

5.9 The dynamic-programming point of view

Here we shall adopt a Bayesian view of the problem, in a way similar to Pronzato and Zhigljavsky (1993) and Section 3 of Wynn and Zhigljavsky (1993), and assume that the number N of function evaluations to be performed is fixed *a priori*. We also assume x^* to have a prior distribution uniform on $\mathcal{J}_0 = [0,1]$. We restrict our attention to second-order algorithms, which are based on the comparison between two values of $f(\cdot)$ evaluated in the current uncertainty interval. A similar approach could be considered for first-order line-search algorithms. Throughout the section, $f(\cdot)$ is assumed to be symmetric around x^*.

In previous algorithms, the rule for deletion is based on the uni-

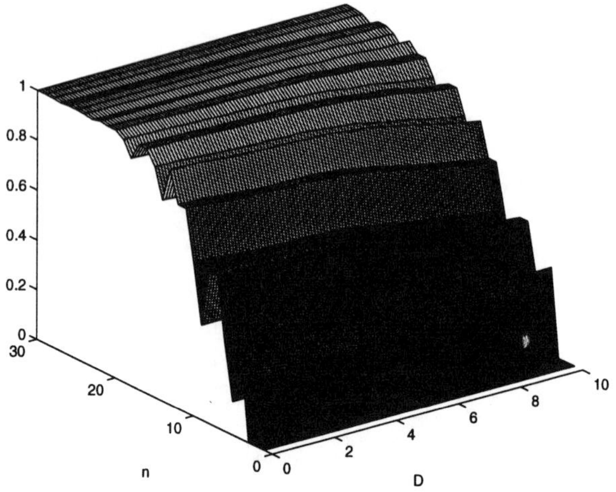

Figure 5.19 *Evolution of P_n^F as a function of n and D for the GS4 algorithm*

extremality assumption, the symmetry of $f(\cdot)$ only being used to update the support of the distribution of x^*. Deleting parts of the interval where the probability density of x^* is zero leads to other procedures. Starting from a uniform distribution on $[0,1]$, the distribution of x^* remains uniform on all intervals $\mathcal{J}_k$, $\mathcal{J}_k = [A_k, B_k]$ after k evaluations of $f(\cdot)$. Optimal procedures for fixed N are constructed by consideration of the backward-in-time problem. The last interval $\mathcal{J}_{N-1}$ is divided into equal parts when the last two evaluations are performed symmetrically, and this step is average optimal; that is, the expected length of the last interval is then minimal. The same argument applies for previous iterations, which yields an average-optimal procedure. It can be initialised for instance by two evaluations in $\mathcal{J}_1 = [0,1]$ at (1/3, 2/3), which produces a cycle with two states. The final length L_N is equal to $L_0/2^{N-1}$, and the rate $(\prod_{i=0}^{N-1} r_i)^{1/N}$ is equal to $(1/2)^{(N-1)/N}$. Since the reduction rate r_i is constant, this approach is also optimal in the minimax sense.

Consider again a deletion rule based only on the uni-extremality assumption. This always yields larger intervals than the ones obtained previously, so that $(1/2)^{(N-1)/N}$ is a lower bound for the rate of any second-order algorithm. Procedures that reach this bound asymptotically have been considered in Wynn and Zhigl-

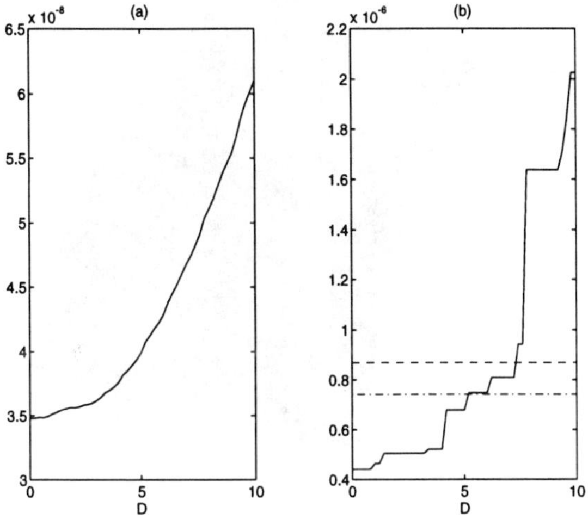

Figure 5.20 *Evolution of* $\mathbb{E}L_{30}$ *(left) and* $\mathbb{M}L_{30}$ *(right) as functions of D for the window algorithm. The dashed and dash-dotted horizontal lines on the right respectively correspond to* φ^{29} *and* $1/F_{31}$*, that is, to the GS and Fibonacci algorithms*

javsky (1995) for locally symmetric functions, and ϵ-optimal algorithms with a finite number of states are presented in Example 5.10.

In what follows, we wish to minimise $\mathbb{E}_{x^*}\{L_N(S_N)\}$ with respect to the strategy S_N, with N fixed. The restriction to the class of quasi-symmetrical strategies (*i.e.*, to procedures for which all observations are taken symmetrically except the last two ones) has already been considered in Pronzato and Zhigljavsky (1993). Replacing average optimality by minimax optimality leads to the Fibonacci algorithm (Kiefer, 1953, 1957). Minimax-optimal and quasi-symmetrical average-optimal strategies tend asymptotically (when $N \to \infty$) to the GS algorithm.

The distribution of x^* in each interval $\mathcal{J}_k$, conditional on the information available in the interval, is uniform on the *optimistic interval* $[\alpha_k, \beta_k]$ obtained by using the optimistic rule defined in Section 5.4.1. Note the difference with Pronzato and Zhigljavsky (1993), where for simplicity the distribution was taken uniform over

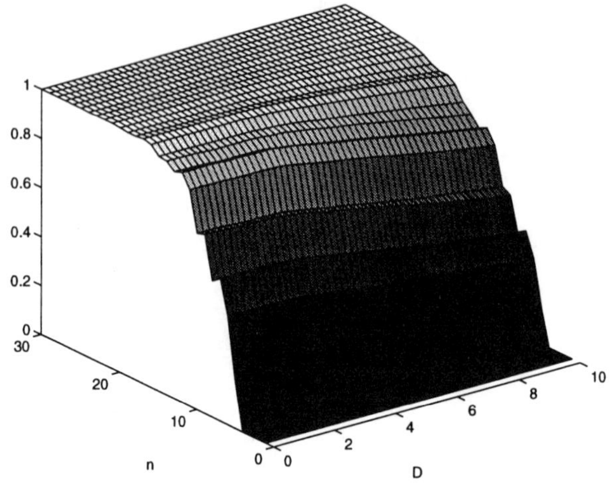

Figure 5.21 *Evolution of P_n^F as a function of n and D for the window algorithm*

$\mathcal{J}_k$; that is, the assumption of symmetry for $f(\cdot)$ was not taken into account when propagating the conditional density of x^*.

Define e_k, e'_k as in Section 3.2.3; see (3.9). The minimisation of $\mathbf{E}_{x^*}\{L_N(S_N)\}$ corresponds to the solution of the following dynamic-programming problem

$$\begin{aligned}\min_{e_1,e'_1} \quad & \mathbf{E}_{x^*|\alpha_1,\beta_1}\{r_2(e_1,e'_1) \times \min_{e'_2} \mathbf{E}_{x^*|\alpha_2,\beta_2}\{r_3(e_2,e'_2)\ldots \\ & \times \min_{e'_{N-2}} \mathbf{E}_{x^*|\alpha_{N-2},\beta_{N-2}}\{r_{N-1}(e_{N-2},e'_{N-2}) \\ & \times \min_{e_N} \mathbf{E}_{x^*|\alpha_{N-1},\beta_{N-1}}\{r_N(e_{N-1},e_N)\}\}\ldots\}\}.\end{aligned}$$

Define $l_k(e_k,\alpha_k,\beta_k)$ as

$$l_k(e_k,\alpha_k,\beta_k) = \mathbf{E}_{x^*|\alpha_k,\beta_k}\{r_k(e_k,e'_k)\mathbf{E}_{x^*|\alpha_{k+1},\beta_{k+1}} \\ \times\{r_{k+1}(e_{k+1},e'_{k+1})\ldots \times \mathbf{E}_{x^*|\alpha_{N-1},\beta_{N-1}}\{r_N(e_{N-1},e_N)\}\ldots\}\},$$

where we omit the dependence of l_k over the e'_m's, $k \leq m < N$.

Since the distribution of x^* is uniform on $[\alpha_k, \beta_k]$, we get

$$l_k(e_k, \alpha_k, \beta_k) = \begin{cases} \frac{\beta_k - c_k}{\beta_k - \alpha_k}(1 - e_k) l_{k+1}\left(\frac{e'_k - e_k}{1 - e_k}, \frac{c_k - e_k}{1 - e_k}, \frac{\beta_k - e_k}{1 - e_k}\right) \\ \quad + \frac{c_k - \alpha_k}{\beta_k - \alpha_k} e'_k l_{k+1}\left(\frac{e_k}{e'_k}, \frac{\alpha_k}{e'_k}, \frac{c_k}{e'_k}\right) \quad \text{if } e'_k > e_k\,, \\ \\ \frac{\beta_k - c_k}{\beta_k - \alpha_k}(1 - e'_k) l_{k+1}\left(\frac{e_k - e'_k}{1 - e'_k}, \frac{c_k - e'_k}{1 - e'_k}, \frac{\beta_k - e'_k}{1 - e'_k}\right) \\ \quad + \frac{c_k - \alpha_k}{\beta_k - \alpha_k} e_k l_{k+1}\left(\frac{e'_k}{e_k}, \frac{\alpha_k}{e_k}, \frac{c_k}{e_k}\right) \quad \text{otherwise}\,, \end{cases}$$

with $c_k = (e_k + e'_k)/2$. The expected length $\mathbf{E}_{x^*}\{L_N\} = l_1(e_1, 0, 1)$ can thus be calculated, at least in principle, for any choice of the e'_k's.

We follow again a backward-in-time approach, and consider the interval $\mathcal{J}_{N-1}$ in order to express $e_N = e'_{N-1}$ as a function of e_{N-1}, α_{N-1} and β_{N-1}. Assuming first that $e_N \le e_{N-1}$, e_N must be chosen so as to minimise

$$l_N^0(e_N) = (1 - e_N)\left(\beta_{N-1} - \frac{e_N + e_{N-1}}{2}\right) + e_{N-1}\left(\frac{e_N + e_{N-1}}{2} - \alpha_{N-1}\right),$$

with the constraints

$$\alpha_{N-1} \le \frac{e_N + e_{N-1}}{2} \le \beta_{N-1} \text{ and } 0 \le e_N\,,$$

or equivalently

$$e_N \in \mathcal{J}_{N-1}^0 = [\max(0, 2\alpha_{N-1} - e_{N-1}), \min(e_{N-1}, 2\beta_{N-1} - e_{N-1})]\,.$$

The optimum e_N^0 is then either $1/2 + \beta_{N-1} - e_{N-1}$ if this point lies in $\mathcal{J}_{N-1}^0$, or one of the bounds of $\mathcal{J}_{N-1}^0$. Assuming now that $e_N \ge e_{N-1}$, e_N must be chosen so as to minimise

$$l_N^1(e_N) = (1 - e_{N-1})\left(\beta_{N-1} - \frac{e_N + e_{N-1}}{2}\right) + e_N\left(\frac{e_N + e_{N-1}}{2} - \alpha_{N-1}\right),$$

with the admissible interval

$$\mathcal{J}_{N-1}^1 = [\max(e_{N-1}, 2\alpha_{N-1} - e_{N-1}), \min(2\beta_{N-1} - e_{N-1}, 1)]\,.$$

The optimum e_N^1 is then either $1/2 + \alpha_{N-1} - e_{N-1}$ if this point lies in $\mathcal{J}_{N-1}^1$, or one of the bounds of $\mathcal{J}_{N-1}^1$. Finally, the optimal choice for e_N is obtained by comparing the criteria $l_N^0(e_N^0)$ and $l_N^1(e_N^1)$. This gives

$$l_{N-1}(e_{N-1},\alpha_{N-1},\beta_{N-1}) =$$

$$\begin{cases} e_{N-1} - \frac{(2\beta_{N-1}-1)^2}{8(\beta_{N-1}-\alpha_{N-1})} & \text{if } e_N = e_N^0 \\ 1 - e_{N-1} - \frac{(2\alpha_{N-1}-1)^2}{8(\beta_{N-1}-\alpha_{N-1})} & \text{if } e_N = e_N^1 \\ \frac{2e_{N-1}^2 - e_{N-1}(1+\alpha_{N-1}+\beta_{N-1})+\beta_{N-1}}{\beta_{N-1}-\alpha_{N-1}} & \text{if } e_N = e_{N-1} \\ e_{N-1} & \text{if } e_N = 2\beta_{N-1} - e_{N-1} \\ 1 - e_{N-1} & \text{if } e_N = 2\alpha_{N-1} - e_{N-1} \end{cases}$$

Although the same approach could theoretically be used to express the optimal choice for e'_{N-2} as a function of e_{N-2}, α_{N-2} and β_{N-2}, it is clearly untractable. In what follows we shall therefore restrict our attention to fixed control policies, that is, to cases where e'_k is a fixed function of e_k, α_k and β_k,

$$e'_k = g(e_k, \alpha_k, \beta_k)\,.$$

The only degree of freedom then lies in the choice of the initial point e_1, and replacing e'_k by its expression in $l_k(e_k, \alpha_k, \beta_k)$ we can compute $l_1(e_1, 0, 1)$. Symmetry within the optimistic interval $[\alpha_k, \beta_k]$ is a desirable property for the reduction rate to asymptotically converge to $1/2$, so that we consider the control function defined by

$$e'_k = \alpha_k + \beta_k - e_k\,.$$

Figure 5.22 presents $l_1(e_1, 0, 1)$ as a function of e_1 for various values of N when e_N is chosen optimally as a function of e_{N-1}, α_{N-1} and β_{N-1}. Contrarily to the case considered in Pronzato and Zhigljavsky (1993), no analytical expression is obtained for $l_1(e_1, 0, 1)$ around the optimum.

Simplifying the choice of the last point according to the suboptimal policy $e_N = e_{N-1}$, the following expressions can be derived

$$l_1(e_1, 0, 1) = \begin{cases} 2e_1^2 - 2e_1 + 1 & \text{if } N = 2 \\ 6e_1^2 - 8e_1 + 3 & \text{if } N = 3 \\ 14e_1^2 - 18e_1 + 6 & \text{if } N = 4 \\ 26e_1^2 - 71/2e_1 + 49/4 & \text{if } N = 5 \end{cases}$$

which are valid around their associated optimum starting point (in $[1/2, 1]$). These optimum starting points are thus

$$e_1 = \begin{cases} 1/2 & \text{if } N = 2 \\ 2/3 & \text{if } N = 3 \\ 9/14 & \text{if } N = 4 \\ 71/104 & \text{if } N = 5 \end{cases}$$

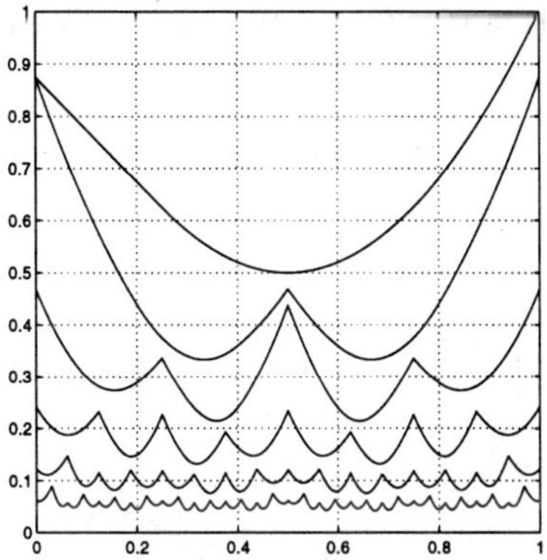

Figure 5.22 $l_1(e_1, 0, 1)$ *as a function of* e_1 *when* e_N *is chosen optimally, with* N *varying from 2 (top curve) to 7 (bottom curve)*

The evolution of $l_1(e_1, 0, 1)$ as a function of e_1 for various values of N is given on Figure 5.23, to be compared with Figure 5.22. The value obtained when $e_1 = 0$ is easily calculated,

$$l_1(0, 0, 1) = \frac{1}{2^{N-3}} \, .$$

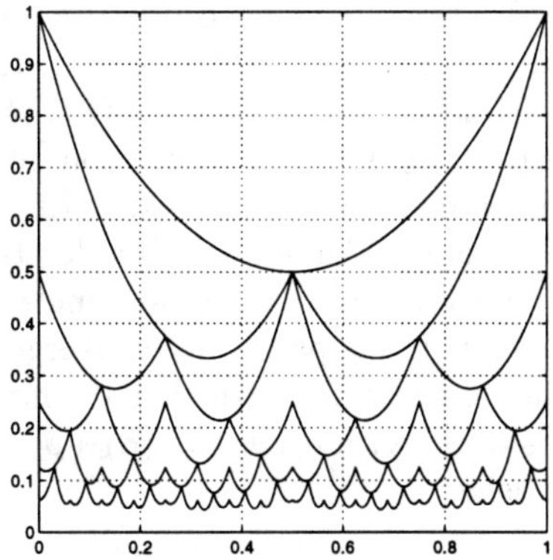

Figure 5.23 $l_1(e_1, 0, 1)$ *as a function of* e_1 *when* $e_N = e_{N-1}$*, with* N *varying from 2 (top curve) to 7 (bottom curve)*

Finally, we consider the case where the last point e_N is also chosen according to a symmetrical policy, *i.e.*, $e_N = \alpha_{N-1} + \beta_{N-1} -$

e_{N-1}. The evolution of $l_1(e_1, 0, 1)$ as a function of e_1 is presented on Figure 5.24.

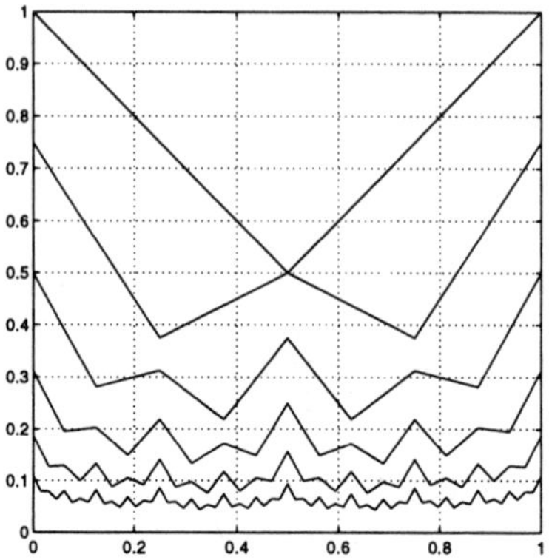

Figure 5.24 $l_1(e_1, 0, 1)$ *as a function of* e_1 *when* $e_N = \alpha_{N-1} + \beta_{N-1} - e_{N-1}$, *with* N *varying from 2 (top curve) to 7 (bottom curve)*

The optimal starting point can then be obtained analytically,

$$e_1 = \frac{2}{3}\left(1 - \left(-\frac{1}{2}\right)^N\right),$$

which tends to 2/3 when N tends to infinity. This policy can be compared with the Fibonacci algorithm (Kiefer, 1953, 1957), where at each step e'_k is taken equal to $1 - e_k$ (*i.e.*, symmetrical with respect to the center $(a_k + b_k)/2$ of the non-normalised interval $[a_k, b_k]$). For both policies, when starting with the optimal value for e_1, each e_k is an optimal starting point for the sequence with $N + 1 - k$ evaluations allowed. When symmetry with respect to $(a_k + b_k)/2$ is considered, the procedure converges to the GS algorithm, which is the unique symmetrical algorithm with two states. When symmetry is taken with respect to $(\alpha_k + \beta_k)/2$, the procedure asymptotically satisfies

$$e_k = \frac{\alpha_k + \beta_k}{2} \pm \frac{\beta_k - \alpha_k}{6},$$

which corresponds to the unique algorithm with two states in the optimistic interval $[\alpha_k, \beta_k]$.

CHAPTER 6

Ellipsoid algorithms

6.1 Volume-optimal outer and inner ellipsoids

We have already encountered the construction of inner and outer ellipsoids with respectively maximum and minimum volumes for the ellipsoid algorithms of Section 2.3.4. Their renormalisation, using the unit ball $\mathcal{B}(\mathbf{0}, 1)$ as a base region $\mathcal{R}$, has been considered in Sections 3.3.1 and 3.3.3. Volume-optimal inner and outer ellipsoids have many more applications than the construction of optimisation algorithms (*e.g.*, in control, system identification, design centering, robust estimation, etc.) and one can refer to the survey paper Walter and Pronzato (1994) for a list of references. Here, we pay special attention to the relation with experimental design problems, which also forms a potential field of application for ellipsoid algorithms.

6.1.1 Minimum-volume outer ellipsoids

An algorithm has been presented in Section 2.3.4 for the construction of the minimum-volume ellipsoid containing a truncated ellipsoid. This type of algorithm can be used to construct a recursive (inequalities are treated one-by-one) outer approximation for a polytope, see Bland, Goldfarb and Todd (1981), and is classically used to construct feasible parameter sets in system identification with linear models and bounded noise; see Pronzato and Walter (1994). We consider here the construction of the minimum-volume outer ellipsoid $\mathcal{E}_o^*(\mathcal{C})$ for a polytope $\mathcal{C}$ defined by its vertices. Most of the results we present are valid when $\mathcal{C}$ is a compact set with non-empty interior. From the Loewner-Behrend Theorem, see Berger (1979), we know that the minimum-volume ellipsoid containing $\mathcal{C}$ is then unique.

The determination of $\mathcal{E}_o^*(\mathcal{C})$ is simpler when the centre of the ellipsoid is fixed. We shall denote by $\overline{\mathcal{E}}_o^*(\mathcal{C})$ such an optimal ellipsoid with fixed centre. In fact, one can show that the determination of $\mathcal{E}_o^*(\mathcal{C})$ in dimension d corresponds to the determination of an ellipsoid $\overline{\mathcal{E}}_o^*(\mathcal{C}')$ in dimension $d+1$. We give two proofs. The first is geometric (Shor and Berezovski, 1992; Khachiyan and Todd, 1993); the second relies on a duality with experimental design problems.

Centred and non-centered ellipsoids

We assume that the origin $\mathbf{0}$ belongs to $\mathcal{C}$. Consider the polytope $\mathcal{C}' \subset \mathbb{R}^{d+1}$ defined by

$$\mathcal{C}' = \left\{ \begin{pmatrix} \boldsymbol{x} \\ 1 \end{pmatrix}, \ \boldsymbol{x} \in \mathcal{C} \right\},$$

and the hyperplane

$$\mathcal{H} = \{\boldsymbol{x}' \in \mathbb{R}^{d+1} \ / \ [\boldsymbol{x}']_{d+1} = 1\}.$$

Let $\mathcal{E}'$ denote an ellipsoid of $\mathbb{R}^{d+1}$, centred at the origin,

$$\mathcal{E}' = \{\boldsymbol{x}' \in \mathbb{R}^{d+1} \ / \ \boldsymbol{x}'^T \boldsymbol{M}' \boldsymbol{x}' \leq 1\},$$

and define $\mathcal{E} = \mathcal{E}' \cap \mathcal{H}$, with

$$\mathcal{E} = \{\boldsymbol{x} \in \mathbb{R}^d \ / \ (\boldsymbol{x} - \boldsymbol{c})^T \boldsymbol{M} (\boldsymbol{x} - \boldsymbol{c}) \leq 1\}.$$

Then, $\mathcal{C}' \subset \mathcal{E}'$ is equivalent to $\mathcal{C} \subset \mathcal{E}$, and $\boldsymbol{M}'$, $\boldsymbol{M}$ and $\boldsymbol{c}$ are related by

$$\begin{aligned} \boldsymbol{M}' &= \begin{pmatrix} \tilde{\boldsymbol{M}} & \boldsymbol{r} \\ \boldsymbol{r}^T & q \end{pmatrix}, \\ \boldsymbol{c} &= -\tilde{\boldsymbol{M}}^{-1} \boldsymbol{r}, \\ \boldsymbol{M} &= \frac{\tilde{\boldsymbol{M}}}{1 - q + \boldsymbol{r}^T \tilde{\boldsymbol{M}}^{-1} \boldsymbol{r}}. \end{aligned}$$

Defining the height of $\mathcal{E}'$ as

$$h = \max\{[\boldsymbol{x}']_{d+1}, \ \boldsymbol{x}' \in \mathcal{E}'\},$$

one has $h = [\boldsymbol{M}'^{-1}]_{d+1,d+1} = (q - \boldsymbol{r}^T \tilde{\boldsymbol{M}}^{-1} \boldsymbol{r})^{-1/2}$, and

$$\mathtt{vol}(\mathcal{E}') = \mathtt{vol}(\mathcal{E}) \frac{\vartheta_{d+1}}{\vartheta_d} \frac{h^{d+1}}{(h^2 - 1)^{d/2}},$$

with ϑ_d the volume of the d-dimensional unit ball. When $\mathcal{E}' = \overline{\mathcal{E}}_o^*(\mathcal{C}')$, then $h = \sqrt{d+1}$ and $\mathcal{E} = \mathcal{E}' \cap \mathcal{H}$ coincides with $\mathcal{E}_o^*(\mathcal{C})$; see Shor and Berezovski (1992), Khachiyan and Todd (1993).

Connection with experiment design

Consider a problem of parameter estimation for a linear regression model with unknown parameters $\bar{\boldsymbol{\theta}} \in I\!R^d$ and scalar observations $\boldsymbol{y} = \{y(1), \dots, y(n)\}$; that is,

$$y(k) = \boldsymbol{r}^T(k)\bar{\boldsymbol{\theta}} + \epsilon(k).$$

The errors $\epsilon(k)$ are assumed to be i.i.d. with zero mean and known variance. We assume that the regressors $\boldsymbol{r}(k)$, $k = 1, \dots, n$, belong to a given compact set $\mathcal{X} \subset I\!R^d$. The experimental design problem consists in choosing the regressors so that the estimation of $\boldsymbol{\theta}$ from the data $\boldsymbol{y}$ is as precise as possible. We consider least-squares estimation; that is, we estimate $\boldsymbol{\theta}$ by

$$\hat{\boldsymbol{\theta}}(\boldsymbol{y}) = \arg\min_{\boldsymbol{\theta} \in I\!R^d} \sum_{k=1}^{n} [y(k) - \boldsymbol{r}^T(k)\boldsymbol{\theta}]^2 = (\boldsymbol{R}^T\boldsymbol{R})^{-1}\boldsymbol{R}^T\boldsymbol{y},$$

with

$$\boldsymbol{R} = \begin{pmatrix} \boldsymbol{r}^T(1) \\ \vdots \\ \boldsymbol{r}^T(n) \end{pmatrix}.$$

A natural measure of the precision of the estimation is a scalar function of the *information matrix*, given by

$$\boldsymbol{M}(\Xi) = \boldsymbol{R}^T\boldsymbol{R} = \sum_{k=1}^{n} \boldsymbol{r}(k)\boldsymbol{r}^T(k),$$

the inverse of which is proportional to the covariance matrix of $\hat{\boldsymbol{\theta}}(\boldsymbol{y})$. Here Ξ denotes the set of regressors used in the experiment, $\Xi = \{\boldsymbol{r}(1), \dots, \boldsymbol{r}(n)\}$. A classical criterion for designing the experiment, that is, for choosing Ξ, is the determinant of $\boldsymbol{M}(\Xi)$, and Ξ_D is said D-optimal if and only if

$$\Xi_D = \arg\max_{\Xi} \det \boldsymbol{M}(\Xi);$$

see, *e.g.*, Fedorov (1972). If replicating regressors in Ξ is allowed, one can write

$$\boldsymbol{J}(\Xi) = \frac{\boldsymbol{M}(\Xi)}{n} = \sum_{k=1}^{m} \frac{n_i}{n} \boldsymbol{r}(i)\boldsymbol{r}^T(i),$$

with $\sum_{i=1}^{m} n_i = n$, where n_i is the number of replications of the i-th regressor. The ratio n_i/n can then be interpreted as the percentage of experimental effort devoted to this i-th regressor. This can be generalized (Kiefer and Wolfowitz, 1959) to the notion of design measure.

Consider a normalised measure ξ on $\mathcal{X}$, $\int \xi(d\boldsymbol{x}) = 1$, and define the *information matrix per sample* as

$$\boldsymbol{J}(\xi) = \int \boldsymbol{x}\boldsymbol{x}^T \xi(d\boldsymbol{x}) .$$

The experimental design problem then amounts to the determination of a measure ξ that maximises a scalar function $\psi(\cdot)$ of $\boldsymbol{J}(\xi)$, usually such that $\psi[\boldsymbol{J}(\xi)]$ is concave with respect to ξ. For instance, a D-optimal design measure ξ_D maximises $\det \boldsymbol{J}(\xi)$. From the Caratheodory Theorem, an optimal design measure with finite support always exists; see, *e.g.*, Fedorov (1972). Note that at least d support points are required to make $\boldsymbol{J}(\xi)$ nonsingular. The next theorem states that the determination of ξ_D is dual to the solution of an optimal ellipsoidal problem.

Theorem 6.1 *The determination of a normalised measure ξ_D on $\mathcal{X}$ that maximises $\det \boldsymbol{J}(\xi)$ is the dual problem of the determination of the minimum-volume ellipsoid $\overline{\mathcal{E}}_o^*(\mathcal{X})$ centred at the origin and containing $\mathcal{X}$. The support points of ξ_D are the points of contact of $\overline{\mathcal{E}}_o^*(\mathcal{X})$ with $\mathcal{X}$, and $\overline{\mathcal{E}}_o^*(\mathcal{X})$ is given by*

$$\overline{\mathcal{E}}_o^*(\mathcal{X}) = \{\boldsymbol{x} \in \mathbb{R}^d \ / \ \boldsymbol{x}^T \boldsymbol{J}^{-1}(\xi_D)\boldsymbol{x} \leq d\} .$$

The proof is a direct application of Lagrangian theory; see Sibson (1972), Silvey (1980). The Kiefer and Wolfowitz equivalence Theorem (1960) gives further properties for ξ_D.

Theorem 6.2 *The following statements are equivalent:*
(i) ξ_D is D-optimal (it maximises $\det \boldsymbol{J}(\xi)$),
(ii) $\xi_D = \arg\min_\xi \max_{\boldsymbol{x}\in\mathcal{X}} d(\boldsymbol{x}, \xi)$,
(iii) $\max_{\boldsymbol{x}\in\mathcal{X}} d(\boldsymbol{x}, \xi_D) = d$,
with

$$d(\boldsymbol{x}, \xi) = \boldsymbol{x}^T \boldsymbol{J}^{-1}(\xi)\boldsymbol{x} . \tag{6.1}$$

An additional property is that a D-optimal measure with at most $d(d+1)/2$ support points always exists. When the number of support points is d, they all receive the same weight $1/d$. Note that although ξ_D is not necessarily unique, $\boldsymbol{J}(\xi_D)$ is, which makes $\overline{\mathcal{E}}_o^*(\mathcal{X})$ unique, in agreement with the Loewner-Behrend Theorem.

In the case of ellipsoids with free centre, one gets the following theorem; see Pronzato and Walter (1994).

Theorem 6.3 *The determination of the minimum-volume ellipsoid $\mathcal{E}_o^*(\mathcal{X})$ containing $\mathcal{X}$ is dual to the determination of the normalised measure ξ_V that maximises* $\det \boldsymbol{J}_c(\xi)$, *with*

$$\boldsymbol{J}_c(\xi) = \int \boldsymbol{x}\boldsymbol{x}^T \xi(d\boldsymbol{x}) - \boldsymbol{c}(\xi)\boldsymbol{c}^T(\xi),$$

where

$$\boldsymbol{c}(\xi) = \int \boldsymbol{x}\ \xi(d\boldsymbol{x}).$$

The support points of ξ_V are the points of contact of $\mathcal{E}_o^*(\mathcal{X})$ with $\mathcal{X}$, and $\mathcal{E}_o^*(\mathcal{X})$ is given by

$$\mathcal{E}_o^*(\mathcal{X}) = \{\boldsymbol{x} \in \mathbb{R}^d \ / \ [\boldsymbol{x} - \boldsymbol{c}(\xi_V)]^T \boldsymbol{J}_c^{-1}(\xi_V)[\boldsymbol{x} - \boldsymbol{c}(\xi_V)] \leq d\}.$$

A direct extension of the Equivalence Theorem 6.2 is as follows.

Theorem 6.4 *The following statements are equivalent:*
(i) ξ_V *maximises* $\det \boldsymbol{J}_c(\xi)$,
(ii) $\xi_V = \arg\min_\xi \max_{\boldsymbol{x}\in\mathcal{X}} d_c(\boldsymbol{x}, \xi)$,
(iii) $\max_{\boldsymbol{x}\in\mathcal{X}} d_c(\boldsymbol{x}, \xi_V) = d$,
with

$$d_c(\boldsymbol{x}, \xi) = [\boldsymbol{x} - \boldsymbol{c}(\xi)]^T \boldsymbol{J}_c^{-1}(\xi)[\boldsymbol{x} - \boldsymbol{c}(\xi)]. \qquad (6.2)$$

The number of support points of ξ_V is now at least $d+1$ and can always be bounded by $d(d+3)/2$. The fact that the determination of a minimum-volume ellipsoid with free centre in $\mathbb{R}^d$ corresponds to the determination of a minimum-volume ellipsoid with fixed centre in $\mathbb{R}^{d+1}$ can be seen as follows.

From Theorem 6.3, $\mathcal{E}_o^*(\mathcal{X})$ is obtained by the determination of ξ_V that maximises $\det \boldsymbol{J}_c(\xi)$, which can be written as

$$\det \boldsymbol{J}_c(\xi) = \det \begin{pmatrix} \boldsymbol{J}(\xi) & \boldsymbol{c}(\xi) \\ \boldsymbol{c}^T(\xi) & 1 \end{pmatrix} = \det \boldsymbol{J}'(\xi),$$

with

$$\boldsymbol{J}'(\xi) = \int \boldsymbol{z}\boldsymbol{z}^T \xi(d\boldsymbol{z}), \text{ and } \boldsymbol{z} = \begin{pmatrix} \boldsymbol{x} \\ 1 \end{pmatrix} \in \mathbb{R}^{d+1}. \qquad (6.3)$$

Determining ξ_V is thus equivalent to determining ξ'_D that maximises $\det \boldsymbol{J}'(\xi)$, that is, to solving a D-optimal design problem with regressors in $\mathbb{R}^{d+1}$. Note that we keep the same notation for the design measure on points $\boldsymbol{x} \in \mathbb{R}^d$ and $\boldsymbol{z} \in \mathbb{R}^{d+1}$ since there is a one-to-one correspondence between them.

6.1.2 Algorithms for minimum-volume outer ellipsoids

From Theorem 6.1, the following type of algorithm, classical in experiment design, can be used to construct minimum-volume outer ellipsoids with predefined centre.

Algorithm 1. (Steepest ascent for the determination of $\overline{\mathcal{E}}_o^*(\mathcal{C})$.)

Step (i) Choose $0 < \delta << 1$ and take a non-degenerate discrete distribution ξ^1 on $\mathcal{C}$. Set $k = 1$.

Step (ii) Compute $\boldsymbol{x}^+ = \arg\max_{\boldsymbol{x} \in \mathcal{C}} d(\boldsymbol{x}, \xi^k)$, where $d(\boldsymbol{x}, \xi)$ is given by (6.1).

Step (iii) If $d(\boldsymbol{x}^+, \xi^k) < \delta$ stop; otherwise

$$\xi^{k+1} = (1-\alpha)\xi^k + \alpha \xi_{\boldsymbol{x}^+}\,, \quad k \to k+1\,,$$

where $\xi_{\boldsymbol{x}}$ is the discrete distribution with weight 1 at $\boldsymbol{x}$, and

$$\alpha = \alpha_k^* = \frac{d(\boldsymbol{x}^+, \xi^k) - d}{d(d(\boldsymbol{x}^+, \xi^k) - 1)}\,, \tag{6.4}$$

or

$$\alpha = \alpha_k\,, \quad \text{with} \lim_{k\to\infty} \alpha_k = 0\,, \ \sum_{n=1}^{\infty} \alpha_n = \infty\,. \tag{6.5}$$

The choice (6.4) corresponds to the maximisation of $\det \boldsymbol{J}(\xi^{k+1})$ with respect to α; see Fedorov (1972). The convergence of $\det \boldsymbol{J}(\xi^k)$ is then monotonous, which is not the case when (6.5) is used. Note that, although we look for an ellipsoid with minimum volume, the sequence of volumes is increasing! Convergence for the case where (6.5) is used is proved in Wynn (1970).

From Theorem 6.3, the following algorithm can be used to determine $\mathcal{E}_o^*(\mathcal{C})$ with free centre.

Algorithm 2. (Steepest ascent for the determination of $\mathcal{E}_o^*(\mathcal{C})$.)

Step (i) Choose $0 < \delta << 1$ and take a non-degenerate discrete distribution ξ^1 on $\mathcal{C}$. Set $k = 1$.

Step (ii) Compute $\boldsymbol{x}^+ = \arg\max_{\boldsymbol{x} \in \mathcal{C}} d_c(\boldsymbol{x}, \xi^k)$, where $d_c(\boldsymbol{x}, \xi)$ is given by (6.2).

Step (iii) If $d_c(\boldsymbol{x}^+, \xi^k) < \delta$ stop; otherwise

$$\xi^{k+1} = (1-\alpha)\xi^k + \alpha \xi_{\boldsymbol{x}^+}\,, \quad k \to k+1\,,$$

where $\xi_{\boldsymbol{x}}$ is the discrete distribution with weight 1 at $\boldsymbol{x}$, and

$$\alpha = \alpha_k^* = \frac{d_c(\boldsymbol{x}^+, \xi^k) - d}{(d+1)d_c(\boldsymbol{x}^+, \xi^k)}, \tag{6.6}$$

or $\alpha = \alpha_k$ given by (6.5).

Note that

$$\alpha_k^* = \frac{d'(\boldsymbol{x}^+, \xi^k) - (d+1)}{(d+1)(d'(\boldsymbol{x}^+, \xi^k) - 1)},$$

with

$$d'(\boldsymbol{x}, \xi) = (\boldsymbol{x}^T \; 1)^T \boldsymbol{J}'^{-1}(\xi) \begin{pmatrix} \boldsymbol{x} \\ 1 \end{pmatrix} = d_c(\boldsymbol{x}, \xi) + 1, \tag{6.7}$$

and $\boldsymbol{J}'(\xi)$ given by (6.3), so that the choice $\alpha = \alpha_k^*$ corresponds to the Fedorov algorithm. Again, the sequence of volumes is monotonously increasing. Various methods have been suggested in the literature to accelerate the convergence of steepest ascent algorithms; see Fedorov and Hackl (1997), Walter and Pronzato (1997) for references.

When $\mathcal{C}$ is a polytope, the support points of the optimal design measure are vertices of $\mathcal{C}$. Let $\boldsymbol{x}_i$, $i = 1, \ldots, m$ denote these vertices. Only the associated weights z_i have then to be determined, with $z_i \geq 0, i = 1, \ldots, m$, and $\sum_{i=1}^m z_i = 1$. The following recursion gives a sequence of ellipsoid that converges to $\overline{\mathcal{E}}_o^*(\mathcal{C})$; see Titterington (1976), Pázman (1986):

$$z_i^{k+1} = z_i^k \frac{d(\boldsymbol{x}_i, \xi^k)}{d}, \quad i = 1, \ldots, m,$$

where $d(\boldsymbol{x}, \xi)$ is given by (6.1) and ξ^k is the discrete distribution with weight z_i on the support point $\boldsymbol{x}_i$. The recursion is initialised by a distribution that puts a strictly positive weight at each vertex of $\mathcal{C}$.

A straightforward extension to the determination of $\mathcal{E}_o^*(\mathcal{C})$ is

$$z_i^{k+1} = z_i^k \frac{d'(\boldsymbol{x}_i, \xi^k)}{d+1}, \quad i = 1, \ldots, m, \tag{6.8}$$

with $d'(\boldsymbol{x}, \xi)$ given by (6.7). We conjecture that for $d > 1$ the following recursion also gives a sequence of ellipsoids converging to $\mathcal{E}_o^*(\mathcal{C})$:

$$z_i^{k+1} = z_i^k \frac{d_c(\boldsymbol{x}_i, \xi^k)}{d}, \quad i = 1, \ldots, m, \tag{6.9}$$

with $d_c(\boldsymbol{x}, \xi)$ given by (6.2). As reported in Titterington (1978), numerical examples show that convergence is faster than with the recursion (6.8).

6.1.3 Maximum-volume inner ellipsoids

From the Danzer-Zagustin Theorem, see Danzer, Grünbaum and Klee (1963), there exists a unique maximum-volume ellipsoid $\mathcal{E}_i^*(\mathcal{C})$ contained in a convex set $\mathcal{C}$. Again, the determination of this ellipsoid is much simpler when its centre is fixed, and we shall denote by $\overline{\mathcal{E}}_i^*(\mathcal{C})$ such a volume-optimal inner ellipsoid with fixed centre. This centre can always be taken at the origin, and we shall consider the set $\mathcal{K}_o(\mathbb{R}^d)$ of convex compact subsets $\mathcal{C}$ of $\mathbb{R}^d$ that contain the origin $\mathbf{0}$ in their interior. Consider the polar transformation $\Phi(\cdot)$ given by

$$\begin{array}{rcl} \mathcal{K}_o(\mathbb{R}^d) & \mapsto & \mathcal{K}_o(\mathbb{R}^d) \\ \mathcal{C} & \rightarrow & \Phi(\mathcal{C}) = \{\boldsymbol{z} \in \mathbb{R}^d \ / \ \boldsymbol{z}^T\boldsymbol{x} \leq 1, \forall \boldsymbol{x} \in \mathcal{C}\}\,. \end{array}$$

It satisfies the following properties.

Theorem 6.5

(i) $\Phi(\cdot)$ defines a relation of duality in the following sense:

(a) $\forall \mathcal{C} \in \mathcal{K}_o(\mathbb{R}^d), \Phi[\Phi(\mathcal{C})] = \mathcal{C}$,

(b) for any given polytope $\mathcal{C}$ in $\mathcal{K}_o(\mathbb{R}^d)$, $\Phi(\cdot)$ defines a one-to-one relation between the q-faces of $\mathcal{C}$ and the $(d-q-1)$-faces of $\Phi(\mathcal{C})$ (for a polytope in $\mathbb{R}^d$, a vertex is a 0-face, an edge is a 1-face, a hyperplane is a $(d-1)$-face, a q-face is a face of a $(q+1)$-face).

(ii) $\Phi(\cdot)$ satisfies

$$\forall(\mathcal{C}_1, \mathcal{C}_2) \in \mathcal{K}_o(\mathbb{R}^d)^2, \mathcal{C}_1 \subset \mathcal{C}_2 \Leftrightarrow \Phi(\mathcal{C}_2) \subset \Phi(\mathcal{C}_1)\,.$$

(iii) When $\mathcal{C}$ is a polytope of $\mathcal{K}_o(\mathbb{R}^p)$ defined by

$$\mathcal{C} = \{\boldsymbol{x} \in \mathbb{R}^d \ / \ \boldsymbol{a}_i^T\boldsymbol{x} \leq b_i, i = 1, \ldots, m\}\,, \qquad (6.10)$$

the vertices of $\Phi(\mathcal{C})$ belong to the set $\{\boldsymbol{a}_i/b_i, i = 1, \ldots, m\}$.

(iv) If $\mathbf{0} \in \mathcal{E}(\mathbf{c}, \boldsymbol{A})$, then $\Phi[\mathcal{E}(\mathbf{c}, \boldsymbol{A})] = \mathcal{E}(\mathbf{c}', \boldsymbol{A}')$, with

$$\left\{ \begin{array}{rcl} \mathbf{c}' & = & -\frac{\boldsymbol{A}^{-1}\mathbf{c}}{1-\mathbf{c}^T\boldsymbol{A}^{-1}\mathbf{c}}\,, \\ \boldsymbol{A}' & = & \frac{1}{1-\mathbf{c}^T\boldsymbol{A}^{-1}\mathbf{c}}(\boldsymbol{A}^{-1} + \frac{\boldsymbol{A}^{-1}\mathbf{c}\mathbf{c}^T\boldsymbol{A}^{-1}}{1-\mathbf{c}^T\boldsymbol{A}^{-1}\mathbf{c}})\,. \end{array} \right.$$

Results (i) and (ii) are classical; see Berger (1979). The proof of (iii) and (iv) is straightforward; see Pronzato and Walter (1996). We have in particular $\Phi[\mathcal{E}(\mathbf{0}, \boldsymbol{A})] = \mathcal{E}(\mathbf{0}, \boldsymbol{A}^{-1})$, which gives

$$\texttt{vol}\{\Phi[\mathcal{E}(\mathbf{0}, \boldsymbol{A})]\}\ \texttt{vol}[\mathcal{E}(\mathbf{0}, \boldsymbol{A})] = \vartheta_d^2 . \tag{6.11}$$

Consider the determination of the maximum-volume ellipsoid $\overline{\mathcal{E}}_i^*(\mathcal{C})$ with fixed centre. This centre can be taken as the origin with no loss of generality. From Theorem 6.5 (ii), $\mathcal{E}(\mathbf{0}, \boldsymbol{A}) \subset \mathcal{C}$ if and only if $\Phi(\mathcal{C}) \subset \Phi[\mathcal{E}(\mathbf{0}, \boldsymbol{A})]$, and from (6.11) maximizing $\texttt{vol}[\mathcal{E}(\mathbf{0}, \boldsymbol{A})]$ is equivalent to minimizing $\texttt{vol}\{\Phi[\mathcal{E}(\mathbf{0}, \boldsymbol{A})]\}$. The determination of $\overline{\mathcal{E}}_i^*(\mathcal{C})$ is thus equivalent to that of $\overline{\mathcal{E}}_o^*[\Phi(\mathcal{C})]$. Note that if $\mathcal{C}$ is defined by inequalities of the form (6.10), then from Theorem 6.5 (iii), $\Phi(\mathcal{C})$ is defined by its vertices. The determination of $\overline{\mathcal{E}}_o^*[\Phi(\mathcal{C})]$ can thus be performed with the algorithms presented in Section 6.1.2.

We do not have any similar equivalence between inner and outer ellipsoid problems when the centre of the ellipsoid is free. However, given an ellipsoid $\mathcal{E}(\boldsymbol{c}, \boldsymbol{A}) \subset \mathcal{C}$, we can increase its volume as shown by the following theorem.

Theorem 6.6 *Let $\mathcal{E}(\boldsymbol{c}, \boldsymbol{A})$ be an inner ellipsoid for the convex set $\mathcal{C}$, and define $\tau_{\boldsymbol{c}}(\cdot)$ as the translation in $\mathbb{R}^d$: $\boldsymbol{z} \to \tau_{\boldsymbol{c}}(\boldsymbol{z}) = \boldsymbol{z} - \boldsymbol{c}$. One has*

$$\tau_{-\boldsymbol{c}}[\Phi(\mathcal{E}_o^*\{\Phi[\tau_{\boldsymbol{c}}(\mathcal{C})]\})] \subset \mathcal{C} , \tag{6.12}$$

and

$$\texttt{vol}\{\tau_{-\boldsymbol{c}}[\Phi(\mathcal{E}_o^*\{\Phi[\tau_{\boldsymbol{c}}(\mathcal{C})]\})]\} \geq \texttt{vol}[\mathcal{E}(\boldsymbol{c}, \boldsymbol{A})] . \tag{6.13}$$

Proof. One has

$$\mathcal{E}_o^*\{\Phi[\tau_{\boldsymbol{c}}(\mathcal{C})]\} \supset \Phi[\tau_{\boldsymbol{c}}(\mathcal{C})] ,$$

and thus from Theorem 6.5 (i–ii),

$$\Phi(\mathcal{E}_o^*\{\Phi[\tau_{\boldsymbol{c}}(\mathcal{C})]\}) \subset \tau_{\boldsymbol{c}}(\mathcal{C}) ,$$

which gives (6.12). Moreover,

$$\texttt{vol}(\mathcal{E}_o^*\{\Phi[\tau_{\boldsymbol{c}}(\mathcal{C})]\}) \leq \texttt{vol}(\overline{\mathcal{E}}_o^*\{\Phi[\tau_{\boldsymbol{c}}(\mathcal{C})]\}) ,$$

since the centre of $\mathcal{E}_o^*\{\Phi[\tau_{\boldsymbol{c}}(\mathcal{C})]\}$ is chosen optimally. Denote $\mathcal{E}(\boldsymbol{d}, \boldsymbol{B}) = \mathcal{E}_o^*\{\Phi[\tau_{\boldsymbol{c}}(\mathcal{C})]\}$. From Theorem 6.5 (iv), one has

$$\texttt{vol}[\mathcal{E}(\boldsymbol{d}, \boldsymbol{B})]\texttt{vol}\{\Phi[\mathcal{E}(\boldsymbol{d}, \boldsymbol{B})]\} = \vartheta_d^2 \frac{1}{(1 - \boldsymbol{d}^T \boldsymbol{B}^{-1} \boldsymbol{d})^{(d+1)/2}} \geq \vartheta_d^2 ,$$

so that

$$\begin{aligned}\mathtt{vol}\{\tau_{-c}[\Phi(\mathcal{E}_o^*\{\Phi[\tau_c(\mathcal{C})]\})]\} &\geq \frac{\vartheta_d^2}{\mathtt{vol}(\mathcal{E}_o^*\{\Phi[\tau_c(\mathcal{C})]\})}\\ &\geq \frac{\vartheta_d^2}{\mathtt{vol}(\overline{\mathcal{E}}_o^*\{\Phi[\tau_c(\mathcal{C})]\})}\\ &= \mathtt{vol}\{\overline{\mathcal{E}}_i^*[\tau_c(\mathcal{C})]\}\\ &\geq \mathtt{vol}[\mathcal{E}(c, A)]\,.\end{aligned}$$

■

It is shown in Pronzato and Walter (1996) that equality in (6.13) implies that $\mathcal{E}(c, A) = \mathcal{E}_i^*(\mathcal{C})$, and that the sequence of ellipsoids generated by repeating the transformation

$$\mathcal{E}_k = \mathcal{E}(c, A) \to \mathcal{E}_{k+1} = \tau_{-c}[\Phi(\mathcal{E}_o^*\{\Phi[\tau_c(\mathcal{C})]\})]$$

converges to $\mathcal{E}_i^*(\mathcal{C})$. This forms a first algorithm for the construction of $\mathcal{E}_i^*(\mathcal{C})$: we solve a sequence of subproblems corresponding to the determination of a minimum-volume outer ellipsoid. Another algorithm is given in Khachiyan and Todd (1993), which amounts to solving a sequence of subproblems corresponding to the determination of the maximal paraboloid contained in a cone. A detailed study of the polynomial complexity of the algorithm is presented. Shor and Berezovski (1992) suggest an algorithm of the subgradient type; see Shor (1985). Other methods are presented in the next section. They are based on the outer-ellipsoid algorithm of Section 2.3.4.

6.1.4 Algorithms for constructing inner ellipsoids

We consider a polytope $\mathcal{C}$ defined by linear inequalities (6.10), and wish to determine the maximum-volume ellipsoid inscribed in $\mathcal{C}$. Without any loss of generality, we assume that the origin $\mathbf{0}$ belongs to the interior of $\mathcal{C}$, so that the b_i's are positive. We can thus renormalise the inequalities, and define $\mathcal{C}$ as

$$\mathcal{C} = \{x \in \mathbb{R}^d \ / \ a_i^T x \leq 1\,,\ i = 1, \ldots, m\}\,. \qquad (6.14)$$

Let the optimal ellipsoid $\mathcal{E}_i^*(\mathcal{C})$ be defined by (A, c):

$$\mathcal{E}_i^*(\mathcal{C}) = \mathcal{E}(c, A) = \{x \in \mathbb{R}^d \ / \ (x - c)^T A^{-1}(x - c) \leq 1\}\,;$$

A and c should then be such that $-\log\det A$ is minimum, with the constraints $a_i^T c + \sqrt{a_i^T A a_i} \leq 1$, $i = 1, \ldots, m$. Starting with an ellipsoid not contained in $\mathcal{C}$, we use the outer-ellipsoid algorithm of Section 2.3.4 to construct a sequence of ellipsoids converging to an ellipsoid contained in $\mathcal{C}$.

Algorithm 1.

Suppose that one constraint, say that defined by i_k^*, is violated by the ellipsoid $\mathcal{E}(\boldsymbol{x}^{(k)}, \boldsymbol{A}_k)$. Consider the *homothetic ellipsoid* $\mathcal{E}(\boldsymbol{x}^{(k)}, \boldsymbol{B}_k)$, with $\boldsymbol{B}_k = d^2\boldsymbol{A}_k$ and denote

$$\alpha_i^k = \frac{\boldsymbol{a}_i^T\boldsymbol{x}^{(k)} - 1}{\sqrt{\boldsymbol{a}_i^T\boldsymbol{B}_k\boldsymbol{a}_i}} \tag{6.15}$$

the depth of the cut of $\mathcal{E}(\boldsymbol{x}^{(k)}, \boldsymbol{B}_k)$ by the i-th constraint $\boldsymbol{a}_i^T\boldsymbol{x} \leq 1$. Then, $\boldsymbol{a}_{i_k^*}^T\boldsymbol{x}^{(k)} + \sqrt{\boldsymbol{a}_{i_k^*}^T\boldsymbol{A}_k\boldsymbol{a}_{i_k^*}} > 1$ is equivalent to $\alpha_{i_k^*} = \alpha_{i_k^*}^k > -1/d$. One can thus apply one iteration of the form (2.22) to obtain the minimum-volume ellipsoid $\mathcal{E}(\boldsymbol{x}^{(k+1)}, \boldsymbol{B}_{k+1})$ containing the part of the ellipsoid $\mathcal{E}(\boldsymbol{x}^{(k)}, \boldsymbol{B}_k)$ which satisfies the constraint:

$$\left\{\begin{array}{lcl} \boldsymbol{x}^{(k+1)} & = & \boldsymbol{x}^{(k)} - \rho\dfrac{\boldsymbol{B}_k\boldsymbol{a}_{i_k^*}}{\sqrt{\boldsymbol{a}_{i_k^*}^T\boldsymbol{B}_k\boldsymbol{a}_{i_k^*}}}, \\ \boldsymbol{B}_{k+1} & = & \sigma\left(\boldsymbol{B}_k - \tau\dfrac{\boldsymbol{B}_k\boldsymbol{a}_{i_k^*}\boldsymbol{a}_{i_k^*}^T\boldsymbol{B}_k}{\boldsymbol{a}_{i_k^*}^T\boldsymbol{B}_k\boldsymbol{a}_{i_k^*}}\right), \end{array}\right. \tag{6.16}$$

with

$$\rho = \frac{1+d\alpha}{d+1}, \quad \sigma = \frac{d^2(1-\alpha^2)}{d^2-1}, \quad \tau = \frac{2(1+d\alpha)}{(d+1)(\alpha+1)}, \tag{6.17}$$

and $\alpha = \alpha_{i_k^*}$.

The ellipsoid $\mathcal{E}(\boldsymbol{x}^{(k+1)}, \boldsymbol{B}_{k+1})$ has a smaller volume than $\mathcal{E}(\boldsymbol{B}_k, \boldsymbol{x}^{(k)})$:

$$\frac{\texttt{vol}[\mathcal{E}(\boldsymbol{x}^{(k+1)}, \boldsymbol{B}_{k+1})]}{\texttt{vol}[\mathcal{E}(\boldsymbol{x}^{(k)}, \boldsymbol{B}_k)]} = \left(\frac{d^2(1-\alpha_{i_k^*}^2)}{d^2-1}\right)^{(d-1)/2}\frac{d(1-\alpha_{i_k^*})}{d+1} < 1\,. \tag{6.18}$$

The homothetic ellipsoid $\mathcal{E}(\boldsymbol{x}^{(k+1)}, \boldsymbol{A}_{k+1})$, with $\boldsymbol{A}_{k+1} = \boldsymbol{B}_{k+1}/d^2$, has thus a smaller volume than $\mathcal{E}(\boldsymbol{x}^{(k)}, \boldsymbol{A}_k)$. One can easily check that $\alpha_{i_k^*}^{k+1} = -1/d$, so that the ellipsoid $\mathcal{E}(\boldsymbol{x}^{(k+1)}, \boldsymbol{A}_{k+1})$ is tangent to the constraint used for the cut at iteration k; see Figure 6.1 which shows an example of a polytope $\mathcal{C}$, together with two homothetic ellipsoids $\mathcal{E}(\boldsymbol{x}^{(k)}, \boldsymbol{A}_k)$ (dotted line) and $\mathcal{E}(\boldsymbol{x}^{(k)}, d^2\boldsymbol{A}_k)$ (dashed line).

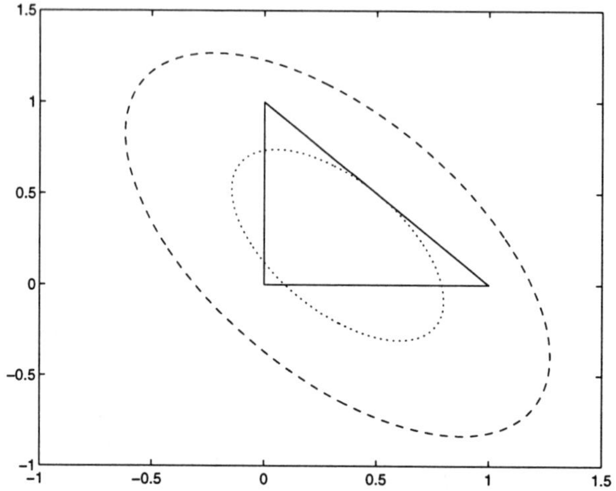

Figure 6.1 *An example of polytope with two homothetic ellipsoids generated by Algorithm 1*

One can show that the sequence of ellipsoids generated by repeating the procedure above converges to an ellipsoid contained in $\mathcal{C}$.

Theorem 6.7 *Consider a bounded polytope $\mathcal{C}$ defined by (6.14). Take an initial ellipsoid $\mathcal{E}(\boldsymbol{x}^{(0)}, \boldsymbol{A}_0)$ such that the interior of $\mathcal{C} \cap \mathcal{E}(\boldsymbol{x}^{(0)}, d^2\boldsymbol{A}_0)$ is not empty. Consider the sequence of ellipsoids $\mathcal{E}(\boldsymbol{x}^{(k)}, d^2\boldsymbol{A}_k)$ generated by algorithm (6.16), where the deepest possible cut is used at each iteration, that is, $\alpha_{i_k^*} = \bar{\alpha}_k$, with*

$$\bar{\alpha}_k = \max_{i=1,\dots,m} \alpha_i^k \,, \tag{6.19}$$

where α_i^k is given by (6.15) for $\boldsymbol{B}_k = d^2\boldsymbol{A}_k$. This sequence converges to an ellipsoid contained in $\mathcal{C}_\delta$:

$$\forall \delta > 0\,,\ \exists K \text{ such that, } \forall k > K\,,\ \mathcal{E}(\boldsymbol{x}^{(k)}, \boldsymbol{A}_k) \subset \mathcal{C}_\delta\,, \tag{6.20}$$

with

$$\mathcal{C}_\delta = \{\boldsymbol{x} \in I\!R^d \ / \ \boldsymbol{a}_i^T \boldsymbol{x} \le 1 + \delta\,,\ i = 1, \dots, m\}\,.$$

Proof. The result is proved if $\mathcal{E}(\boldsymbol{x}^{(0)}, \boldsymbol{A}_0) \subset \mathcal{C}$, and we thus assume that it is not the case. Since $\mathbf{int}[\mathcal{C} \cap \mathcal{E}(\boldsymbol{x}^{(0)}, d^2\boldsymbol{A}_0)]$ is not empty, $\bar{\alpha}_0 < 1$, and, since $\mathcal{C} \cap \mathcal{E}(\boldsymbol{x}^{(k)}, d^2\boldsymbol{A}_k) \subset \mathcal{E}(\boldsymbol{x}^{(k+1)}, d^2\boldsymbol{A}_{k+1})$, $\bar{\alpha}_k < 1$ for any k. There are then two cases.

(i) The algorithm stops after a finite number of iterations K, that is: $\bar{\alpha}_k \leq -1/d$, $\forall k \geq K$. This means that $\mathcal{E}(\boldsymbol{x}^{(k)}, d^2\boldsymbol{A}_k) \subset \mathcal{C}$ $\forall k \geq K$ and the theorem is proved.

(ii) The algorithm does not stop in a finite number of iterations, that is:

$$\forall k\,,\ -\frac{1}{d} < \bar{\alpha}_k < 1\,.$$

Assume that $\bar{\alpha}_k$ does not converge to $-1/d$. Then, there exists a subsequence $(\bar{\alpha}_{k_i})$ converging to some accumulation point $\alpha_l > -1/d$. Since the deepest possible cut is used at each iteration ($\alpha_{i_k^*} = \bar{\alpha}_k$), according to (6.18) there exists γ such that

$$\frac{\texttt{vol}[\mathcal{E}(\boldsymbol{x}^{(k_{i+1})}, \boldsymbol{A}_{k_{i+1}})]}{\texttt{vol}[\mathcal{E}(\boldsymbol{x}^{(k_i)}, \boldsymbol{A}_{k_i})]} \leq \gamma < 1\,.$$

This implies $\lim_{i\to\infty} \texttt{vol}[\mathcal{E}(\boldsymbol{x}^{(k_i)}, \boldsymbol{A}_{k_i})] = 0$, and $\texttt{vol}[\mathcal{E}(\boldsymbol{x}^{(k)}, \boldsymbol{A}_k)]$ decreases monotonously to zero. This contradicts the fact that $\mathcal{C} \cap \mathcal{E}(\boldsymbol{x}^{(0)}, d^2\boldsymbol{A}_0)$ is included in $\mathcal{E}(\boldsymbol{x}^{(k)}, d^2\boldsymbol{A}_k)$, and therefore $\bar{\alpha}_k$ converges to $-1/d$, which implies (6.20). ■

The cuts tend to be shallower as k increases; that is, $\bar{\alpha}_k$ decreases to $-1/d$. In the limit this method gives a pair of so-called *John's ellipsoids* (John, 1948) that satisfy $\mathcal{E}(\boldsymbol{x}^{(k)}, \boldsymbol{A}_k) \subset \mathcal{C} \subset \mathcal{E}(\boldsymbol{x}^{(k)}, d^2\boldsymbol{A}_k)$. We conjecture that, under fairly general conditions on the initial ellipsoid $\mathcal{E}(\boldsymbol{x}^{(0)}, \boldsymbol{A}_0)$, the sequence of ellipsoids $\mathcal{E}(\boldsymbol{x}^{(k)}, \boldsymbol{A}_k)$ converges to the maximum-volume ellipsoid inscribed in $\mathcal{C}$. This is proved in the case where the number m of constraints defining $\mathcal{C}$ in (6.14) is equal to $d+1$: when $\mathcal{C}$ is a simplex, there exists a unique ellipsoid such that

$$\mathcal{E}(\boldsymbol{x}^{(k)}, \boldsymbol{A}_k) \subset \mathcal{C} \subset \mathcal{E}(\boldsymbol{x}^{(k)}, d^2\boldsymbol{A}_k)\,;$$

see John (1948), and this ellipsoid is such that $\mathcal{E}(\boldsymbol{x}^{(k)}, \boldsymbol{A}_k) = \mathcal{E}_i^*(\mathcal{C})$, $\mathcal{E}(\boldsymbol{x}^{(k)}, d^2\boldsymbol{A}_k) = \mathcal{E}_o^*(\mathcal{C})$. One thus gets the following property.

Theorem 6.8 *Assume that $m = d+1$ in (6.20) and that $\mathcal{C} \subset \mathcal{E}(\boldsymbol{x}^{(0)}, d^2\boldsymbol{A}_0)$. Then the sequence of ellipsoids $\mathcal{E}(\boldsymbol{x}^{(k)}, d^2\boldsymbol{A}_k)$ generated by algorithm (6.16) with $\alpha_{i_k^*} = \bar{\alpha}_k$ converges to $\mathcal{E}_o^*(\mathcal{C})$ and the sequence $\mathcal{E}(\boldsymbol{x}^{(k)}, \boldsymbol{A}_k)$ converges to $\mathcal{E}_i^*(\mathcal{C})$.*

Algorithm 2.

In the method above we applied the iteration (2.22) to an ellipsoid homothetic to the one we are interested in. In fact, we can also apply the same iterations directly to the ellipsoid we want to force into $\mathcal{C}$. Consider iteration k, where the ellipsoid is $\mathcal{E}(\boldsymbol{x}^{(k)}, \boldsymbol{B}_k)$, and assume that a constraint (at least) is violated. Let $\boldsymbol{a}_{i_k^*}$ correspond to the most violated constraint. This gives now $\alpha_{i_k^*} > -1$, since the constraint is violated by $\mathcal{E}(\boldsymbol{x}^{(k)}, \boldsymbol{B}_k)$ and not by $\mathcal{E}(\boldsymbol{x}^{(k)}, \boldsymbol{B}_k/d^2)$ as it was the case for Algorithm 1. We can then cut the ellipsoid by the constraint with a depth α such that the next ellipsoid still violates the constraint, or is tangent to it. This gives the following condition on α:

$$\boldsymbol{a}_{i_k^*}^T \boldsymbol{x}^{(k+1)} + \sqrt{\boldsymbol{a}_{i_k^*}^T \boldsymbol{B}_{k+1} \boldsymbol{a}_{i_k^*}} \geq 1 \,, \tag{6.21}$$

or equivalently $\alpha_{i_k^*}^{k+1} \geq -1$, where $\boldsymbol{x}^{(k+1)}$ and $\boldsymbol{B}_{k+1}$ are given by (6.16,6.17).

At the same time, it is important to prevent the size of the ellipsoid from decreasing too fast (see Lemma 6.1 below), and we impose $\alpha = 0$ (*central cuts*) when the constraint defined by $\boldsymbol{a}_{i_k^*}$ cuts $\mathcal{E}(\boldsymbol{x}^{(k)}, \boldsymbol{B}_k)$ too deeply, that is, when $\alpha_{i_k^*}^k \geq 0$. The condition (6.21) is thus only used for $\alpha_{i_k^*}^k < 0$, which corresponds to $\mathbf{0} \in \mathtt{int}[\mathcal{E}(\boldsymbol{x}^{(k)}, \boldsymbol{B}_k)]$. We can then renormalise $\mathcal{E}(\boldsymbol{x}^{(k)}, \boldsymbol{B}_k)$ in such a way that $\boldsymbol{x}^{(k)} = \mathbf{0}$ and $\boldsymbol{B}_k = \boldsymbol{I}$. This gives $\alpha_i^k = -1/\|\boldsymbol{a}_i\|$, $i = 1, \ldots, m$. Easy calculations show that (6.21) is equivalent to $\alpha \leq l(\alpha_{i_k^*})$ with

$$l(\alpha_{i_k^*}) = \frac{d - 1 + (d+1)\alpha_{i_k^*}}{2d} \,. \tag{6.22}$$

This function is plotted in Figure 6.2. Again, to prevent the size of $\mathcal{E}(\boldsymbol{x}^{(k)}, \boldsymbol{B}_k)$ from decreasing too fast, we impose $\alpha \leq \min[0, l(\alpha_{i_k^*})]$, which gives the dashed area in Figure 6.2 as the admissible region for α. Note that $\alpha = \min(0, \alpha_{i_k^*}/d)$ is suitable and that $\mathtt{vol}[\mathcal{E}(\boldsymbol{x}^{(k+1)}, \boldsymbol{B}_{k+1})] < \mathtt{vol}[\mathcal{E}(\boldsymbol{x}^{(k)}, \boldsymbol{B}_k)]$ only if $\alpha > -1/d$.

The restriction to $\alpha \leq 0$ is active only in the first iterations. The following property shows that it permits to prevent the ellipsoid from vanishing.

Lemma 6.1 *Assume that* $\mathtt{int}[\mathcal{C} \cap \mathcal{E}(\boldsymbol{x}^{(0)}, \boldsymbol{B}_0)]$ *is not empty. Then if* $\alpha \leq 0$ *at each iteration,* $\mathtt{int}[\mathcal{C} \cap \mathcal{E}(\boldsymbol{x}^{(k)}, \boldsymbol{B}_k)]$ *is not empty for any* k.

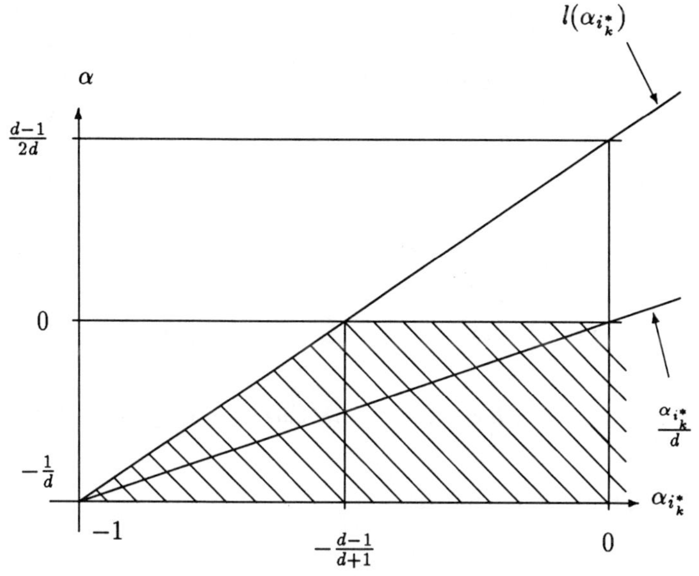

Figure 6.2 *Admissible values of α in Algorithm 2 ($d = 3$)*

Proof. The proof is by induction on k. It is enough to consider the case with the deepest cuts ($\alpha = \min[0, l(\alpha_{i_k^*})]$). Assume that $\mathbf{int}[\mathcal{C} \cap \mathcal{E}(\boldsymbol{x}^{(k)}, \boldsymbol{B}_k)]$ is not empty. Three cases are considered.

(i) $\alpha_{i_k^*} \geq 0$. Then $\mathcal{E}(\boldsymbol{x}^{(k+1)}, \boldsymbol{B}_{k+1})$ contains the half ellipsoid given by $\mathcal{E}(\boldsymbol{x}^{(k)}, \boldsymbol{B}_k) \cap \{\boldsymbol{x} \ / \ \boldsymbol{a}_{i_k^*}^T \boldsymbol{x} \leq \boldsymbol{a}_{i_k^*}^T \boldsymbol{x}^{(k)}\}$, so that $\mathbf{int}[\mathcal{C} \cap \mathcal{E}(\boldsymbol{x}^{(k+1)}, \boldsymbol{B}_{k+1})]$ is not empty.

(ii) $0 > \alpha_{i_k^*} \geq -(d-1)/(d+1)$. Renormalise $\mathcal{E}(\boldsymbol{x}^{(k)}, \boldsymbol{B}_k)$ to the ball $\mathcal{B}(\boldsymbol{x}^{(k)}, 1)$: $\mathcal{C}$ contains $\mathcal{B}(\boldsymbol{x}^{(k)}, -\alpha_{i_k^*})$; see Figure 6.3. Since $\mathcal{B}(\boldsymbol{x}^{(k)}, -\alpha_{i_k^*}) \subset \mathcal{E}(\boldsymbol{x}^{(k+1)}, \boldsymbol{B}_{k+1})$, $\mathbf{int}[\mathcal{C} \cap \mathcal{E}(\boldsymbol{x}^{(k+1)}, \boldsymbol{B}_{k+1})]$ is not empty.

(iii) $-(d-1)/(d+1) > \alpha_{i_k^*} > -1$. Again, one can renormalise $\mathcal{E}(\boldsymbol{x}^{(k)}, \boldsymbol{B}_k)$ to $\mathcal{B}(\boldsymbol{x}^{(k)}, 1)$, and $\mathcal{B}(\boldsymbol{x}^{(k)}, -\alpha_{i_k^*}) \subset \mathcal{C} \cap \mathcal{E}(\boldsymbol{x}^{(k+1)}, \boldsymbol{B}_{k+1})$; see Figure 6.4. ∎

Assume now that at each iteration k we choose α in the dashed

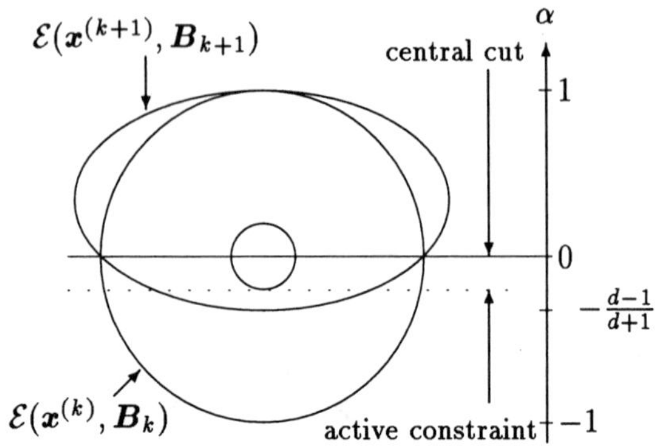

Figure 6.3 *Algorithm 2 with central cuts* ($0 > \alpha_{i_k^*} \geq -(d-1)/(d+1)$)

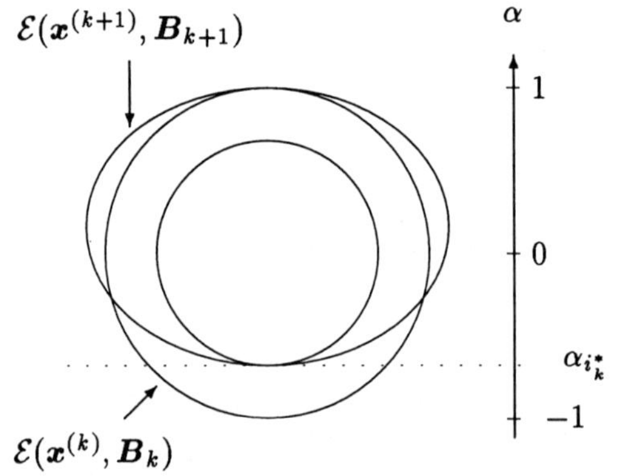

Figure 6.4 *Algorithm 2 with shallow cuts* ($\alpha_{i_k^*} < -(d-1)/(d+1)$)

area of Figure 6.2:

$$\alpha = \min[0, f(\alpha_{i_k^*})], \ f(\alpha_{i_k^*}) \leq l(\alpha_{i_k^*}) \tag{6.23}$$

with the following condition of $f(\cdot)$:

$$\alpha_{i_k^*} > -1 \Rightarrow f(\alpha_{i_k^*}) > -\frac{1}{d}. \tag{6.24}$$

We then have the following convergence property.

Theorem 6.9 *Consider a bounded polytope C defined by (6.14). Take an initial ellipsoid $\mathcal{E}(\boldsymbol{x}^{(0)}, \boldsymbol{B}_0)$ such that* $\mathtt{int}[C \cap \mathcal{E}(\boldsymbol{x}^{(0)}, \boldsymbol{B}_0)]$

is not empty. Consider the sequence of ellipsoids $\mathcal{E}(\boldsymbol{x}^{(k)}, \boldsymbol{B}_k)$ generated by algorithm (6.16,6.17), with α satisfying (6.23,6.24). Then, either $\mathtt{vol}[\mathcal{E}(\boldsymbol{x}^{(k)}, \boldsymbol{B}_k)]$ tends to zero or the sequence of ellipsoids converges to an ellipsoid contained in $\mathcal{C}_\delta$ in the following sense:

$$\forall \delta > 0\,,\ \exists K \text{ such that, } \forall k > K\,,\ \mathcal{E}(\boldsymbol{x}^{(k)}, \boldsymbol{B}_k) \subset \mathcal{C}_\delta\,, \qquad (6.25)$$

with

$$\mathcal{C}_\delta = \{\boldsymbol{x} \in I\!R^d \ /\ \boldsymbol{a}_i^T \boldsymbol{x} \leq 1 + \delta\,,\ i = 1, \ldots, m\}\,.$$

Proof. The result is proved if $\mathcal{E}(\boldsymbol{x}^{(0)}, \boldsymbol{B}_0) \subset \mathcal{C}$, and we thus assume that it is not the case.

From Lemma 6.1, $\mathtt{int}[\mathcal{C} \cap \mathcal{E}(\boldsymbol{x}^{(b)}, \boldsymbol{B}_k)]$ is not empty for all k, and thus $\bar{\alpha}_k < 1$ for all k. There are two cases:

(i) The algorithm stops after a finite number of iterations K, that is: $\bar{\alpha}_k \leq -1$, $\forall k \geq K$. This means that $\mathcal{E}(\boldsymbol{x}^{(k)}, \boldsymbol{B}_k) \subset \mathcal{C}\ \forall k \geq K$ and the theorem is proved.

(ii) The algorithm does not stop in a finite number of iterations, that is:

$$\forall k\,,\ -1 < \bar{\alpha}_k < 1\,.$$

If $\bar{\alpha}_k$ converges to -1, then (6.25) is satisfied. Otherwise, there exists a subsequence $(\bar{\alpha}_{k_i})$ converging to some accumulation point $\alpha_l > -1$, and the cuts for this subsequence thus satisfy $\alpha > -1/d$. Consider then (6.18), with $\alpha_{i_k^*}$ replaced by α:

$$r_k = \frac{\mathtt{vol}[\mathcal{E}(\boldsymbol{x}^{(k_{i+1})}, \boldsymbol{B}_{k_{i+1}})]}{\mathtt{vol}[\mathcal{E}(\boldsymbol{x}^{(k_i)}, \boldsymbol{B}_{k_i})]} \leq \gamma < 1\,. \qquad (6.26)$$

This implies $\lim_{i\to\infty} \mathtt{vol}[\mathcal{E}(\boldsymbol{x}^{(k_i)}, \boldsymbol{B}_{k_i})] = 0$, and $\mathtt{vol}[\mathcal{E}(\boldsymbol{x}^{(k)}, \boldsymbol{B}_k)]$ decreases monotonously to zero. ■

We conjecture that, when properly initialised, Algorithm 2 generates a sequence of ellipsoids $\mathcal{E}(\boldsymbol{x}^{(k)}, \boldsymbol{B}_k)$ that converges to $\mathcal{E}_i^*(\mathcal{C})$. Figure 6.5 gives the ratio (6.26) for Algorithms 1 (full line) and 2 (dashed line) as a function of the violation γ of the constraint when $d = 10$. For Algorithm 1, the depth of the cut is $\alpha = \alpha_{i_k^*} = \gamma/d$; for Algorithm 2 it is $\alpha = \min[0, l(\gamma)]$, with $l(\cdot)$ given by (6.22).

Algorithm 1 is faster when $\alpha_{i_k^*} > 0$, and is thus recommended in the initial phase of the optimisation. Algorithm 2 becomes faster for negative values of $\alpha_{i_k^*}$, which makes it suitable when the ellipsoid becomes tighter. The difference between the convergence rates of the two algorithms increases with d. Note that we used here the limiting value $\alpha = \min[0, l(\alpha_{i_k^*})]$ for Algorithm 2. The choice $\alpha =$

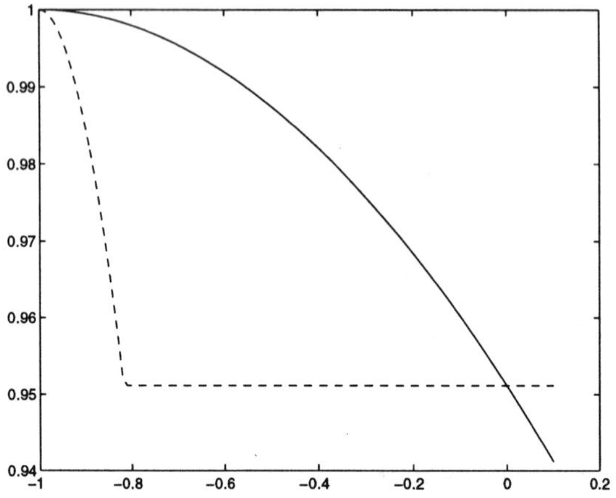

Figure 6.5 *Comparison of the convergence rates of Algorithms 1 (full line) and 2 (dashed line) when* $d = 10$

$\min[0, \alpha_{i_k^*}/d]$; see Figure 6.2, gives the same rate of convergence for the two algorithms for negative values of $\alpha_{i_k^*}$ (although the algorithms are different, since, for instance, $\mathcal{E}(\boldsymbol{x}^{k+1}, \boldsymbol{B}_{k+1})$ is not tangent to the cut used at iteration k for Algorithm 2).

These algorithms can be used for convex and linear programming; see Section 2.3.4. At each iteration of the algorithm the function to be optimised is evaluated at the centre of the maximum-volume ellipsoid contained in a polytope. This ellipsoid need not be evaluated accurately, and a few iterations of Algorithms 1 or 2 may be enough to get a point deep in the polytope. For convex programming, each new iteration of the optimisation algorithm starts by cutting the optimal ellipsoid of previous iteration through its centre, so that Algorithm 2 (with $\alpha_{i_k^*} = 0$) will give better performance than Algorithm 1. In the case of linear programming, the search for the optimal ellipsoid at a iteration $k+1$ can be initialised by reflecting the optimal ellipsoid of iteration k with respect to the hyperplane tangent at the point with the best objective value; see Figure 2.11.

6.2 Ergodic behaviour of the outer-ellipsoid algorithm in linear programming

We consider the outer-ellipsoid algorithm of Sections 2.3.4 and 3.3.1, and study its ergodic behaviour in the 2-dimensional case in some detail.

With the renormalisation and the additional standardisation described in Section 3.3.1, the dynamical system defined by (3.17) can be written as a four-dimensional process. Without loss of generality we may take

$$\boldsymbol{A} = \begin{bmatrix} \cos\theta & \sin\theta \\ \cos\phi & \sin\phi \end{bmatrix}, \qquad \boldsymbol{b} = \begin{bmatrix} \beta_1 \\ \beta_2 \end{bmatrix},$$

and consider $\boldsymbol{z} = (\tan\theta, \tan\phi, \beta_1, \beta_2)$ as the four variables.

6.2.1 The central-cut algorithm

For the central-cut algorithm, one has at all iterations (when the cut is by the objective or the constraint) $\alpha = 0$, and the constants ρ, σ, τ given by (2.23) become: $\rho = 1/3$, $\tau = 2/3$, $\sigma = 4/3$. It can be shown that if we start with $\tan\phi < 0, \tan\theta > 0, \tan\phi \times \tan\theta > -1$ then for all iterations

$$\tan(\phi - \theta) < 0, \quad \tan\phi < 0, \quad \tan\theta > 0, \quad \tan\phi \times \tan\theta > -1\,.$$

It is possible to find the inverse of the iteration and write down the Frobenius-Perron equations, but Figure 6.6 shows that the invariant measure for $(\tan\theta, \tan\phi)$ lives on a fractal, indicating a totally different limiting behaviour from the deep-cut case. The numerical evaluation of the box-counting dimension of the fractal of the four-dimensional process, see, for example, Jensen (1993), gives the estimate $\simeq 2.1$.

6.2.2 The deep-cut algorithm

We consider the case in which α_k is fixed to positive constant α^0 in the objective case and deep cut is performed in the constraint case.

In the objective case $\beta_1 \geq 0$, $\beta_2 \geq 0$ and a typical iteration step

$$\boldsymbol{z} = (\tan\theta, \tan\phi, \beta_1, \beta_2) \mapsto \boldsymbol{z}' = (\tan\theta', \tan\phi', \beta_1', \beta_2')$$

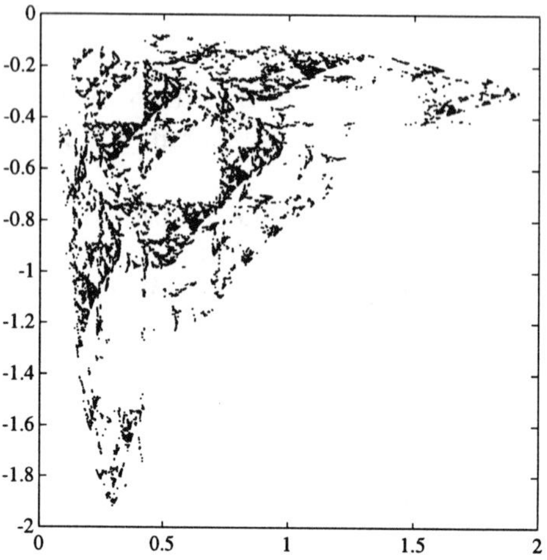

Figure 6.6 *A fractal in the ellipsoid algorithm: sequence of iterates* $(\tan\theta_k, \tan\phi_k)$

is given by

$$\begin{cases} \tan\theta' &= (1-\gamma)\tan\theta \\ \tan\phi' &= (1-\gamma)\tan\phi \\ \beta_1' &= \frac{1}{\sqrt{\sigma}}\frac{\beta_1(1+\tan^2\theta)^{1/2}-\rho\tan\theta}{[1+(1-\gamma)^2\tan^2\theta]^{1/2}} \\ \beta_2' &= \frac{1}{\sqrt{\sigma}}\frac{\beta_2(1+\tan^2\phi)^{1/2}+\rho\tan\phi}{[1+(1-\gamma)^2\tan^2\phi]^{1/2}} \end{cases}$$

where $\gamma = 1-\sqrt{1-\tau}$. Consider the constraint case $\beta_1 < 0,\ \ \beta_2 \geq \beta_1$. Then $\alpha_k = \alpha^A = -\beta_1$ and the iteration is

$$\begin{cases} \tan\theta' &= (1-\gamma)\tan\theta \\ \tan\phi' &= \frac{\tan\phi}{1-\gamma}\left[1-\gamma(2-\gamma)\frac{\tan^2\theta}{1+\tan^2\theta}\right] - \frac{\gamma(2-\gamma)}{1-\gamma}\frac{\tan\theta}{1+\tan^2\theta} \\ \beta_1' &= \frac{\beta_1+\rho}{\sqrt{\sigma}(1-\gamma)} \\ \beta_2' &= \frac{1}{\sqrt{\sigma}}\frac{\beta_2[1+\tan^2(\phi-\theta)]^{1/2}+\rho\,\mathrm{sign}[\tan(\phi-\theta)]}{[(1-\gamma)^2+\tan^2(\phi-\theta)]^{1/2}} \end{cases}$$

For the case $\beta_1 > \beta_2$, $\beta_2 < 0$ we have $\alpha_k - \alpha^R = -\beta_2$ and

$$\begin{cases} \tan\theta' &= \frac{\tan\theta}{1-\gamma}\left[1-\gamma(2-\gamma)\frac{\tan^2\phi}{1+\tan^2\phi}\right] - \frac{\gamma(2-\gamma)}{1-\gamma}\frac{\tan\phi}{1+\tan^2\phi} \\ \tan\phi' &= (1-\gamma)\tan\phi \\ \beta_1' &= \frac{1}{\sqrt{\sigma}}\frac{\beta_1[1+\tan^2(\phi-\theta)]^{1/2}+\rho\,\mathrm{sign}[\tan(\phi-\theta)]}{[(1-\gamma)^2+\tan^2(\phi-\theta)]^{1/2}} \\ \beta_2' &= \frac{\beta_2+\rho}{\sqrt{\sigma}(1-\gamma)} \end{cases}$$

This algorithm exhibits a complex pattern of special periodic attractors which depends on the fixed value α^0 chosen for the objective case; see Figure 6.7. For each attractor, the solution $\boldsymbol{x}_k = \boldsymbol{A}_k^{-1}\boldsymbol{b}_k$ lives on a number of distinct concentric circles. At various critical values of α^0 the number of points in the attractor and the radius of the concentric circles jump; see Figure 6.7. Between these critical points the rate improves with increasing α^0; see Figure 6.8. From the point $\alpha^0 \simeq 0.225$ to $\alpha^0 = 1/2$ there is a 3-point attractor with the rate decreasing smoothly to its minimum 1/3 in the range $0 < \alpha^0 \leq 1/2$. Note the improvement in the convergence rate when going from standard deep cuts, that is, $\alpha^0 = 0$, to deep cuts with $\alpha^0 = 1/2$, especially compared to the marginal improvement obtained when going from central cuts ($r \simeq 0.77$) to standard deep cuts ($r \simeq 0.72$).

After some work the authors have been able to find the attractor at $\alpha^0 = 1/2$. It is given in Table 6.1 and cycles in the order $z_a \rightarrow z_b \rightarrow z_c \rightarrow z_a \ldots$

	β_1	β_2	$\tan\theta$	$\tan\phi$
$\boldsymbol{z}_a$	$-\frac{1}{2}$	$\frac{1}{2}$	$\frac{3\sqrt{3}}{\sqrt{13}}$	$-\frac{1}{\sqrt{3}\sqrt{13}}$
$\boldsymbol{z}_b$	$\frac{1}{2}$	$\frac{\sqrt{3}-1}{\sqrt{10}}$	$\frac{3}{\sqrt{13}}$	$-\frac{3\sqrt{3}}{\sqrt{13}}$
$\boldsymbol{z}_c$	$\frac{\sqrt{3}-1}{\sqrt{10}}$	$-\frac{1}{2}$	$\frac{1}{\sqrt{3}\sqrt{13}}$	$-\frac{3}{\sqrt{13}}$

Table 6.1 *A three-point attractor in the ellipsoid algorithm (d=2)*

6.2.3 *Using super-deep cuts*

Numerical simulations show that even deep cuts can be made deeper by using $\alpha_k = -\lambda\beta_k/||\boldsymbol{s}_k||$, with $\lambda > 1$, when the cut is by the constraint $\boldsymbol{s}_k^T\boldsymbol{x} = \beta_k$. Figure 6.9 presents the renormalised

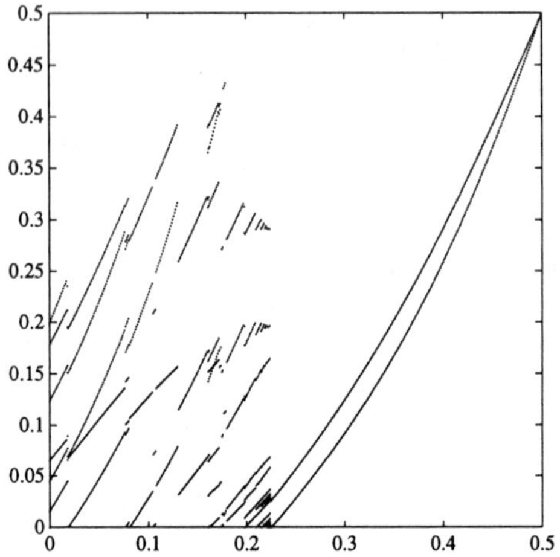

Figure 6.7 *Attractors in the ellipsoid algorithm: depths* α^A *and* α^B *of constraint cuts as function of the fixed depth* α^0 *of the objective cut*

region $\mathcal{B}(\mathbf{0},1)$ and the attractor for the sequence $(\boldsymbol{x}_k = \boldsymbol{A}_k^{-1}\boldsymbol{b}_k)$ in the case $\lambda = 1.121, \alpha^0 = 0.5195$. The attractor is unusual, and illustrates the difficulty of studying the properties of the limiting behaviour of the process.

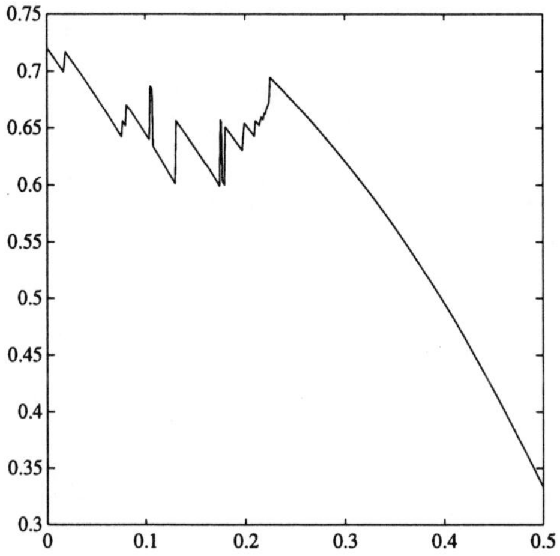

Figure 6.8 *Ergodic convergence rate r as a function of the fixed depth α^0 of the objective cut*

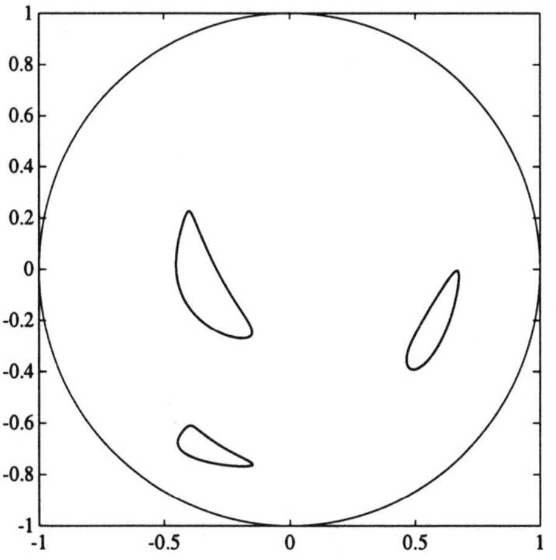

Figure 6.9 *An attractor in the ellipsoid algorithm when using very deep cuts*

CHAPTER 7

Steepest-descent algorithm

In this chapter the behaviour of the standard steepest-descent algorithm for a quadratic function in $\mathbb{R}^d$ is investigated. We show that by rescaling the iterates to remain always on the unit sphere one can reveal special features of this behaviour. The renormalised algorithm is shown to converge to a two-point cycle on the unit circle. The cycle depends on the starting point in a complicated manner (the set of points converging to the same cycle is fractal), but all cycles belong to a particular plane, given by certain eigenvectors of the Hessian matrix of the objective function. We give a proof in the case of standard steepest descent which bears some resemblance to the original Akaike (1959) proof but is hopefully somewhat more transparent. We also give some more precise analysis of the nature of the limit points and of the accelerated algorithm. The stability of the attractor is analysed and the rate of convergence of the algorithm is investigated. It is shown that the worst value of this rate is obtained only for some particular starting points. The introduction of a relaxation coefficient in the steepest-descent algorithm completely changes its behaviour, which may become chaotic. Different attractors are presented. We show that relaxation allows a significantly improved rate of convergence.

One can note that that although neither Akaike nor Forsythe (1968), who also developed the work, stresses the renormalisation aspect, it is inherent in their work. It is probably the first reference in optimisation to the "second order" limiting behaviour inherent in the renormalisation idea of this book.

For a general smooth function $f(\cdot)$ the steepest-descent algorithm is

$$\boldsymbol{x}^{(k+1)} = \boldsymbol{x}^{(k)} - \alpha_k \nabla f(\boldsymbol{x}^{(k)}), \tag{7.1}$$

where

$$\boldsymbol{x}^{(k)} = \left(x_1^{(k)}, \dots, x_d^{(k)}\right)^T$$

is the k-th iterate of the algorithm,

$$\nabla f(\boldsymbol{x}) = \left(\frac{\partial f}{\partial x_1}, \ldots, \frac{\partial f}{\partial x_d}\right)^T$$

is the gradient of $f(\cdot)$, and

$$\alpha_k = \arg\min_{\alpha} f(\boldsymbol{x}^{(k)} - \alpha \nabla f(\boldsymbol{x}^{(k)})) .$$

Let us consider the asymptotic behaviour of this algorithm for a quadratic function in $\mathbb{R}^d$:

$$f(\boldsymbol{x}) = \frac{1}{2}(\boldsymbol{x} - \boldsymbol{x}^*)^T \boldsymbol{A}(\boldsymbol{x} - \boldsymbol{x}^*) ,$$

where $\boldsymbol{x}, \boldsymbol{x}^* \in \mathbb{R}^d$ and $\boldsymbol{A}$ is a positive definite $d \times d$ matrix. This seemingly simple problem has some very complex aspects which are uncovered by the following renormalisation idea. Define

$$\boldsymbol{v}^{(k)} = \frac{\boldsymbol{x}^{(k)} - \boldsymbol{x}^*}{||\boldsymbol{x}^{(k)} - \boldsymbol{x}^*||} , \qquad (7.2)$$

which has the effect of rescaling the iterates to remain always on the unit sphere: $||\boldsymbol{v}|| = 1$.

In Pronzato, Wynn and Zhigljavsky (1995) the authors studied the behaviour of the process $(\boldsymbol{v}^{(k)})$ as $k \to \infty$. Except for pathological cases, in the limit the process was found to belong to the two-dimensional space spanned by the eigenvectors $(\boldsymbol{u}_1, \boldsymbol{u}_d)$ corresponding to the minimal and maximum eigenvalues λ_1 and λ_d of the matrix $\boldsymbol{A}$ under the assumption

$$0 < \lambda_1 < \lambda_2 \leq \ldots \leq \lambda_{d-1} < \lambda_d , \qquad (7.3)$$

where λ_i is the i-th ordered eigenvalue of $\boldsymbol{A}$. Ordering the $\boldsymbol{v}_i^{(k)}$ compatibly with (7.3), one observes $v_i^{(k)} = [\boldsymbol{v}^{(k)}]_i \to 0$ for $i = 2, \ldots, d-1$ as $k \to \infty$, and the algorithm, in its renormalised form, converges to a two-point set on the circle $\{||v_1||^2 + ||v_d||^2 = 1\}$ in the plane spanned by $\boldsymbol{u}_1$ and $\boldsymbol{u}_d$. Numerical simulations show that the same convergence property holds for a convex differentiable function locally quadratic around its minimum $\boldsymbol{x}^*$. It has been known for many years that the asymptotic convergence rate of the algorithm depends on the ratio $\rho = \lambda_d/\lambda_1$ (for instance, see Luenberger, 1973, p. 152). Thus for any $\boldsymbol{x}^{(k)} \in \mathbb{R}^d$

$$f(\boldsymbol{x}^{(k+1)}) \leq \left(\frac{\rho - 1}{\rho + 1}\right)^2 f(\boldsymbol{x}^{(k)}) \qquad (7.4)$$

(see also Section 7.3). However, more precise properties concerning the rate of convergence of the steepest-descent algorithm are difficult to obtain due to the complexity of the behaviour of the renormalised algorithm en route to the limiting circle. This behaviour is fractal in nature, and it still remains open how the limiting asymptotic rate depends on the starting values. It is the worst-case rate, given by (7.4), not the actual rate, which simply depends on λ_1 and λ_d.

In the next section, we show that the algorithm attracts to two "conjugate" points on the circle. We return to the discussion of these conjugate points and the corresponding asymptotic rate of convergence in Sections 7.2 and 7.3 respectively. The behaviour of the steepest-descent algorithm with relaxation is considered in Section 7.4.

7.1 Attraction to a two-dimensional plane

During the course of the research that led to this book and associated papers, we found the following result for the limiting attractor in the quadratic case, using the renormalisation idea. The first proof of the result is due to Akaike (1959) and was generalised by Forsythe (1968). As this book was going to press, we were kindly sent a copy of a technical report by Nocedal, Sartenaer and Zhu (1998) which covers some of the same material from the numerical analysis viewpoint.

Theorem 7.1 *Let the objective function f be*

$$f(\boldsymbol{x}) = \frac{1}{2}(\boldsymbol{x} - \boldsymbol{x}^*)^T \boldsymbol{A}(\boldsymbol{x} - \boldsymbol{x}^*), \tag{7.5}$$

where $\boldsymbol{A}$ is a positive definite matrix with ordered eigenvalues $0 < \lambda_1 < \lambda_2 \leq \ldots \leq \lambda_{d-1} < \lambda_d$. Let $\mathcal{V} = \mathrm{span}(\boldsymbol{u}_1, \boldsymbol{u}_d)$ be the two-dimensional plane generated by the (distinct orthogonal) eigenvectors $\boldsymbol{u}_1$, $\boldsymbol{u}_d$ corresponding to λ_1 and λ_d, respectively. Then for any starting vector $\boldsymbol{x}^{(1)}$ for which

$$\boldsymbol{u}_1^T \boldsymbol{v}^{(1)} \neq 0 \text{ and } \boldsymbol{u}_d^T \boldsymbol{v}^{(1)} \neq 0 \tag{7.6}$$

the steepest descent algorithm attracts to the plane $\mathcal{V}$ in the following sense:

$$\boldsymbol{w}^T \boldsymbol{v}^{(k)} \to 0 \quad k \to \infty$$

for any non-zero vector $\boldsymbol{w} \in \mathcal{V}^{\perp}$, where $\boldsymbol{v}^{(k)}$ is defined by (7.2). Moreover, the sequence $(\boldsymbol{v}^{(k)})$ converges to a two-point cycle.

Proof. For the quadratic function (7.5) the gradient of $f(\cdot)$ at $\boldsymbol{x}^{(k)}$ equals

$$\boldsymbol{g}^{(k)} = \nabla f(\boldsymbol{x}^{(k)}) = \boldsymbol{A}(\boldsymbol{x}^{(k)} - \boldsymbol{x}^*),$$

and

$$\alpha_k = \frac{(\boldsymbol{g}^{(k)})^T \boldsymbol{g}^{(k)}}{(\boldsymbol{g}^{(k)})^T \boldsymbol{A} \boldsymbol{g}^{(k)}}.$$

Therefore, the algorithm (7.1) can be rewritten as

$$\boldsymbol{x}^{(k+1)} = \boldsymbol{x}^{(k)} - \frac{(\boldsymbol{g}^{(k)})^T \boldsymbol{g}^{(k)}}{(\boldsymbol{g}^{(k)})^T \boldsymbol{A} \boldsymbol{g}^{(k)}} \boldsymbol{g}^{(k)}. \tag{7.7}$$

Without loss of generality we can assume that $\boldsymbol{x}^* = 0$ and the matrix $\boldsymbol{A}$ is diagonal: $\boldsymbol{A} = \mathrm{diag}(\lambda_1, \ldots \lambda_d)$ where $0 < \lambda_1 < \lambda_2 \leq \ldots \leq \lambda_{d-1} < \lambda_d$. With this assumption, the iteration of the algorithm (7.7) can be written as

$$\boldsymbol{x}_i^{(k+1)} = \boldsymbol{x}_i^{(k)} - \frac{\sum_{j=1}^d (g_j^{(k)})^2}{\sum_{j=1}^d \lambda_j (g_j^{(k)})^2} g_i^{(k)} \quad \text{for } i = 1, \ldots, d, \tag{7.8}$$

with $g_j^{(k)} = [\boldsymbol{g}^{(k)}]_j$. Multiplying both sides of i-th equation of (7.8) by λ_i we obtain

$$g_i^{(k+1)} = g_i^{(k)} - \frac{\sum_{j=1}^d (g_j^{(k)})^2}{\sum_{j=1}^d \lambda_j (g_j^{(k)})^2} \lambda_i g_i^{(k)} \quad \text{for } i = 1, \ldots, d. \tag{7.9}$$

Introduce now the variables

$$y_i^{(k)} = (g_i^{(k)})^2,$$

and their renormalised versions

$$z_i^{(k)} = \frac{y_i^{(k)}}{\sum_{j=1}^d y_j^{(k)}}. \tag{7.10}$$

Note that $z_i^{(k)} \geq 0$ for $i = 1, \ldots, d$ and $\sum_{j=1}^d z_j^{(k)} = 1$. The equations (7.9) then imply

$$y_i^{(k+1)} = \left(1 - \frac{\lambda_i \sum_{j=1}^d y_j^{(k)}}{\sum_{j=1}^d \lambda_j y_j^{(k)}}\right)^2 y_i^{(k)} \quad \text{for } i = 1, \ldots, d,$$

and

$$z_i^{(k+1)} = z_i^{(k)} \frac{\left(\sum_{j=1}^d \lambda_j z_j^{(k)} - \lambda_i\right)^2}{\sum_{l=1}^d \left(\sum_{j=1}^d \lambda_j z_j^{(k)} - \lambda_l\right)^2 z_l^{(k)}} \quad \text{for } i = 1, \ldots, d. \tag{7.11}$$

Note that the updating formula (7.11) is exactly the same for the weights $z_i^{(k)}$ and $z_j^{(k)}$ corresponding to equal $\lambda_i = \lambda_j$, and therefore these weights can be summed. It follows that a proof based on the assumption that all λ_i are different can easily be extended to the case of ties among $\lambda_2, \ldots, \lambda_{d-1}$. We thus assume that all eigenvalues of $\boldsymbol{A}$ are different: $0 < \lambda_1 < \lambda_2 < \ldots < \lambda_d$. Note also that to prove the convergence to the two-dimensional plane $\mathcal{V}$ for the sequence $(\boldsymbol{x}^{(k)}/||\boldsymbol{x}^{(k)}||)$, it is enough to prove the same property for the sequence $(\boldsymbol{z}^{(k)})$, $k \to \infty$, with $[\boldsymbol{z}^{(k)}]_i = z_i^{(k)}$, $i = 1, \ldots, d$.

We divide the rest of the proof into four parts: we show **(i)** that the sequence (7.11) converges to a two-dimensional plane, **(ii)** that the directions are eventually fixed, and **(iii)** that they fix at $\boldsymbol{u}_1$ and $\boldsymbol{u}_d$. Finally, we show **(iv)** the convergence to a two-point cycle. It is convenient to consider the discrete measures

$$\pi_k = \left\{ \begin{array}{ccc} \lambda_1 & \ldots & \lambda_d \\ z_1^{(k)} & \ldots & z_d^{(k)} \end{array} \right\}$$

which place the weights $z_i^{(k)}$ at the points λ_i, respectively ($i = 1, \ldots, d$). These measures carry all information about the sequence (7.11). Let $\psi(\cdot)$ denote the operator corresponding to the application of (7.11) to a measure π, so that $\pi_{k+1} = \psi(\pi_k)$.

(i) Define the moments

$$\mu_m = \mu_m(\pi_k) = \int \lambda^m d\pi_k(\lambda) = \sum_{j=1}^d \lambda_j^m z_j^{(k)}, \; m = 0, \pm 1, \pm 2, \ldots \tag{7.12}$$

so that $\mu_0 = 1$, $\mu_1 = \sum_{j=1}^d \lambda_j z_j^{(k)}$, etc.. Define also the moment matrices $\boldsymbol{M}_n = \boldsymbol{M}_n(\pi_k)$ by $[\boldsymbol{M}_n]_{j,l} = \mu_{j+l-2}$ $(j, l = 1, \ldots, n)$ so that

$$\boldsymbol{M}_2(\pi_k) = \begin{pmatrix} \mu_0 & \mu_1 \\ \mu_1 & \mu_2 \end{pmatrix} \text{ and } \boldsymbol{M}_3(\pi_k) = \begin{pmatrix} \mu_0 & \mu_1 & \mu_2 \\ \mu_1 & \mu_2 & \mu_3 \\ \mu_2 & \mu_3 & \mu_4 \end{pmatrix}.$$

Then the denominator in the right hand side of (7.11) is

$$D_k = \sum_{l=1}^{d} \left(\sum_{j=1}^{d} \lambda_j z_j^{(k)} - \lambda_l \right)^2 z_l^{(k)} = \mu_2 - \mu_1^2 = \det[\boldsymbol{M}_2(\pi_k)] \,.$$

Using the updating formula (7.11) we have

$$\begin{aligned} D_{k+1} &= \sum_{l=1}^{d} \left(\sum_{j=1}^{d} \lambda_j z_j^{(k+1)} - \lambda_l \right)^2 z_l^{(k+1)} \\ &= \sum_{j=1}^{d} \lambda_j^2 z_j^{(k+1)} - \left(\sum_{j=1}^{d} \lambda_j z_j^{(k+1)} \right)^2 \\ &= \frac{1}{D_k}(\mu_1^2\mu_2 - 2\mu_1\mu_3 + \mu_4) - \frac{1}{D_k^2}(\mu_1^3 - 2\mu_1\mu_2 + \mu_3)^2 \,. \end{aligned}$$

Therefore

$$\begin{aligned} D_{k+1} - D_k &= \frac{2\mu_1\mu_2\mu_3 + \mu_2\mu_4 - \mu_1^2\mu_4 - \mu_3^2 - \mu_2^2}{(\mu_2 - \mu_1^2)^2} \\ &= \frac{\det[\boldsymbol{M}_3(\pi_k)]}{\{\det[\boldsymbol{M}_2(\pi_k)]\}^2} \,. \end{aligned}$$

The conditions (7.6) transfer through the updating formula (7.11) to $z_1^{(k)} > 0$ and $z_d^{(k)} > 0$ for every $k = 1, 2, \ldots$ From simple moment theory we have therefore that the matrix $\boldsymbol{M}_2(\pi_k)$ is positive definite for every $k = 1, 2, \ldots$ which implies $D_k = \det[\boldsymbol{M}_2(\pi_k)] > 0$. Also, $\boldsymbol{M}_3(\pi_k)$ is a non-negative definite moment matrix and $\det[\boldsymbol{M}_3(\pi_k)] \geq 0$. Thus $D_{k+1} - D_k \geq 0$ which means that the sequence (D_k) is monotonously non-decreasing. Since π_k is a probability measure with bounded support, $D_k = \det[\boldsymbol{M}_2(\pi_k)]$ is bounded above by some constant D_* and therefore the sequence (D_k) converges monotonically to a limit and $D_{k+1} - D_k \to 0$ when $k \to \infty$. Moreover,

$$\det[\boldsymbol{M}_3(\pi_k)] = (D_{k+1} - D_k)D_k^2 \leq (D_{k+1} - D_k)D_*^2 \,,$$

so that $\det[\boldsymbol{M}_3(\pi_k)] \to 0$ when $k \to \infty$.

Note that using the Binet-Cauchy lemma

$$D_k = \det[\boldsymbol{M}_2(\pi_k)] \;=\; \sum_{i<j} z_i^{(k)} z_j^{(k)} (\lambda_j - \lambda_i)^2 \geq D_1 > 0 \,,$$

$$\det[\boldsymbol{M}_3(\pi_k)] = \sum_{i<j<l} z_i^{(k)} z_j^{(k)} z_l^{(k)} \times (\lambda_j - \lambda_i)^2 (\lambda_l - \lambda_j)^2 (\lambda_i - \lambda_l)^2 . \quad (7.13)$$

For a fixed iteration k define the pair (i_k, j_k) which achieves $\max_{i<j} z_i^{(k)} z_j^{(k)}$. (If there are several such pairs, take the smallest of them in, say, lexicographical order.) Then for every k we have

$$D_1 \leq D_k \leq z_{i_k}^{(k)} z_{j_k}^{(k)} \sum_{i<j} (\lambda_j - \lambda_i)^2 .$$

Therefore

$$\delta \leq z_{i_k}^{(k)} z_{j_k}^{(k)} , \quad \delta < z_{i_k}^{(k)} < 1 - \delta , \quad \delta < z_{j_k}^{(k)} < 1 - \delta , \qquad (7.14)$$

where

$$\delta = \frac{D_1}{\sum_{i<j} (\lambda_j - \lambda_i)^2} > 0 . \qquad (7.15)$$

From (7.13) we have

$$\det[\boldsymbol{M}_3(\pi_k)] \geq z_{i_k}^{(k)} z_{j_k}^{(k)} (\lambda_{j_k} - \lambda_{i_k})^2 \sum_{i \neq i_k, j_k} z_i^{(k)} (\lambda_i - \lambda_{i_k})^2 (\lambda_i - \lambda_{j_k})^2 .$$

Since all λ_i are distinct, $\det[\boldsymbol{M}_3(\pi_k)] \to 0$ and $z_{i_k}^{(k)} z_{j_k}^{(k)} \geq \delta > 0$, we have

$$\sum_{i \neq i_k, j_k} z_i^{(k)} \to 0, \quad k \to \infty . \qquad (7.16)$$

This finishes part **(i)** of the proof. The interpretation is that although (i_k, j_k) depends on k the total weight associated with all other points tends to zero.

(ii) We will show in addition to (7.16) that there exists an iteration number k_* such that for all $k \geq k_*$ the pair (i_k, j_k) becomes fixed; that is, $(i_k, j_k) = (i_*, j_*)$ for some $1 \leq i_* < j_* \leq d$ and all $k \geq k_*$.

Let

$$\epsilon_* = \frac{\delta D_1}{(\lambda_d - \lambda_1)^2} ,$$

where δ is defined in (7.15). According to (7.16), there exists $k_* = k_*(\epsilon_*) \geq 1$ such that the total weight associated with all other points different from (i_k, j_k) is smaller than ϵ_* for all $k \geq k_*$. Therefore, for $i \notin \{i_k, j_k\}$ and $k \geq k_*$ the updating formula (7.11) gives

$$z_i^{(k+1)} = z_i^{(k)} \frac{\left(\sum_{j=1}^{d} \lambda_j z_j^{(k)} - \lambda_i\right)^2}{D_k} < \frac{\epsilon_*(\lambda_d - \lambda_1)^2}{D_1} = \delta\,.$$

From (7.14), $z_{i_{k+1}}^{(k+1)} > \delta$ and $z_{j_{k+1}}^{(k+1)} > \delta$, and therefore $i \notin \{i_k, j_k\}$ implies $i \notin \{i_{k+1}, j_{k+1}\}$. This proves that $(i_k, j_k) = (i_*, j_*)$ for some $1 \le i_* < j_* \le d$ and all $k \ge k_*$.

(iii) Let us prove that $(i_*, j_*) = (1, d)$. Recall again that the assumption (7.6) and the updating formula (7.11) imply that $z_1^{(k)} > 0$ and $z_d^{(k)} > 0$ for all k.

Assume that $j_* < d$. Denote

$$\epsilon^* = \min\{\epsilon_*,\, \delta \min_{1\le i<j<d}(\lambda_j - \lambda_i)/\lambda_d\}\,,$$

and assume that $k^* = k^*(\epsilon^*) \ge k_*$ is such that $(i_k, j_k) = (i_*, j_*)$ and the total weight associated with all other points different from (i_*, j_*) is smaller than ϵ^* for all $k \ge k^*$. The existence of such a k^* follows from (i) and (ii).

For convenience, rewrite the algorithm (7.11) in the form

$$z_i^{(k+1)} = z_i^{(k)} \frac{(\mu_1 - \lambda_i)^2}{D_k} \quad \text{for } i = 1, \ldots, d\,, \tag{7.17}$$

and note that

$$\begin{aligned}
\mu_1 = \sum_{j=1}^{d} \lambda_j z_j^{(k)} &= \lambda_{i_*} z_{i_*}^{(k)} + \lambda_{j_*} z_{j_*}^{(k)} + \sum_{j \ne i_*, j_*} \lambda_j z_j^{(k)} \\
\le \lambda_{i_*} z_{i_*}^{(k)} + \lambda_{j_*} z_{j_*}^{(k)} + \lambda_d \sum_{j \ne i_*, j_*} z_j^{(k)} &\le \lambda_{i_*} z_{i_*}^{(k)} + \lambda_{j_*} z_{j_*}^{(k)} + \epsilon^* \lambda_d \\
\le \delta\lambda_{i_*} + (1-\delta)\lambda_{j_*} + \epsilon^*\lambda_d &= \lambda_{j_*} + \delta(\lambda_{i_*} - \lambda_{j_*}) + \epsilon^*\lambda_d \\
&\le \lambda_{i_*} - \delta \min_{1\le i<j<d}(\lambda_j - \lambda_i) + \epsilon^*\lambda_d \le \lambda_{j_*}\,.
\end{aligned}$$

Therefore (7.17) implies that for all $k \ge k^*$

$$\frac{z_d^{(k+1)}}{z_d^{(k)}} = \frac{(\lambda_d - \mu_1)^2}{D_k} > \frac{(\lambda_{j_*} - \mu_1)^2}{D_k} = \frac{z_{j_*}^{(k+1)}}{z_{j_*}^{(k)}}\,.$$

We have arrived at a contradiction since the sequence $(z_{j_*}^{(k)})$ is bounded from below by $\delta > 0$ while the sequence $(z_d^{(k)})$ tends to zero. Thus, $j_* = d$ and analogously $i_* = 1$.

(iv) Finally, let D^* denote the limit $\lim_{k\to\infty} D_k$ discussed in **(i)**. There are only two discrete measures π^1, π^2 with nonzero weights on λ_1 and λ_d and such that

$$\det[\boldsymbol{M}_2(\pi^1)] = \det[\boldsymbol{M}_2(\pi^2)] = D^* ,$$

namely

$$\pi^1 = \left\{ \begin{array}{cc} \lambda_1 & \lambda_d \\ p & 1-p \end{array} \right\} , \quad \pi^2 = \left\{ \begin{array}{cc} \lambda_1 & \lambda_d \\ 1-p & p \end{array} \right\} , \tag{7.18}$$

with

$$p = \frac{1}{2} - \sqrt{\frac{1}{4} - \frac{D^*}{(\lambda_d - \lambda_1)^2}} .$$

Note that $\psi(\pi^1) = \pi^2$ and $\psi(\pi^2) = \pi^1$. Therefore, convergence of D_k to D^* implies convergence of π_k to the limiting cycle $\pi^1 \rightarrow \pi^2 \rightarrow \pi^1 \rightarrow \cdots$ ■

Surprisingly, the updating formula (7.11) coincides with (6.9), the Titterington (1978) algorithm for the determination of a minimum-volume ellipsoid, when the λ_i's are interpreted as the support points $\boldsymbol{x}_i$ and $d = 1$ in (6.9), that is, when the ellipsoid degenerates to an interval. Inspection of the above proof shows the non-convergence of the Titterington algorithm in this case. The conjecture of convergence remains in the case of ellipsoids (dimension ≥ 2), for which we have been unable to discover a steepest descent analogue.

Theorem 1 obviously generalizes to the case where (7.6) may not be satisfied. The algorithm then attracts to a two-dimensional plane $\mathcal{V}'$ and $(\boldsymbol{v}^{(k)})$ converges to a two-point cycle. $\mathcal{V}'$ is defined by the eigenvectors $\boldsymbol{u}_i$, $\boldsymbol{u}_j$ associated with the smallest and largest eigenvalues such that $\boldsymbol{u}_i^T \boldsymbol{v}^{(1)} \neq 0$, $\boldsymbol{u}_j^T \boldsymbol{v}^{(1)} \neq 0$. For the sake of simplicity of notations, we shall assume in what follows that (7.6) is satisfied, and therefore $\boldsymbol{u}_i = \boldsymbol{u}_1$, $\boldsymbol{u}_j = \boldsymbol{u}_d$.

Although the limiting behaviour of the algorithm is simple, its behaviour en route to the attractor is fairly complicated and presents a fractal structure. Figure 7.1 shows the projection onto the plane (v_1, v_3) of the region of attraction for $\boldsymbol{v}^{(k)}$ to a small neighbourhood (radius < 0.02) of the point $\boldsymbol{v}^* = (0.9606, 0, 0.2781)$, with $d = 3$, $\lambda_1 = 1$, $\lambda_2 = 2$, $\lambda_3 = 4$. It illustrates the difficulty of predicting the limiting behaviour, and thus the asymptotic rate of convergence, see Section 7.3, as a function of the starting point.

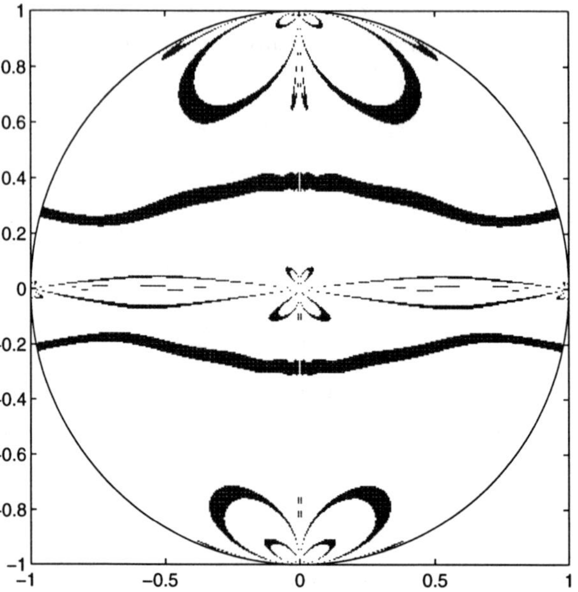

Figure 7.1 *Projection of the region of attraction to a small neighbourhood of the point* $(0.9606, 0, 0.2781)$ *onto the plane* (v_1, v_3) *(*$d = 3, \lambda_1 = 1, \lambda_2 = 2, \lambda_3 = 4$*)*

In Figure 7.2, the grey level of starting points on the unit sphere depends on the limiting value of p in (7.18). Again, this illustrates the complexity of the behaviour of the algorithm on the way to the attractor.

7.2 Stability of attractors

From Theorem 1, the algorithm attracts to two conjugate points on the circle $\{||v_1||^2 + ||v_d||^2 = 1\}$, characterized by the discrete measure (7.18). However, some values of p^* correspond to unstable points. We shall use the following definition of stability; see Hale and Koçak (1991), p. 444.

Definition 7.1 *A fixed point* π *for a mapping* $T(\cdot)$ *is called stable if* $\forall \epsilon > 0$, $\exists \alpha > 0$ *such that* $\forall \pi^0$ *for which* $||\pi^0 - \pi|| < \alpha$, $||T^n(\pi^0) - \pi|| < \epsilon$ *for all* $n > 0$. *A fixed point* π *is unstable if it is not stable.*

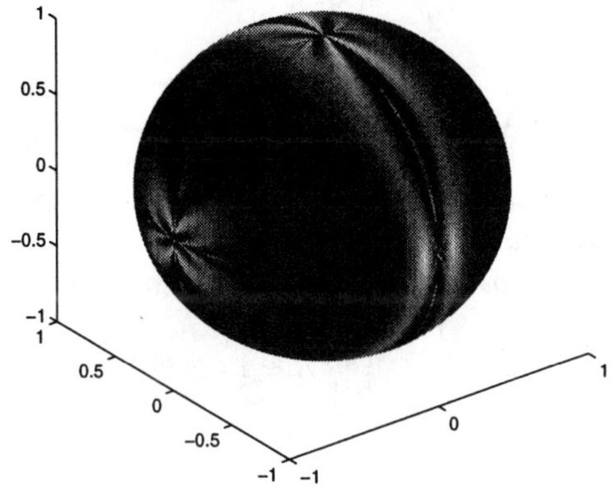

Figure 7.2 *Starting points on the unit sphere coloured as a function of the limiting value of p in (7.18) ($d = 3, \lambda_1 = 1, \lambda_2 = 2, \lambda_3 = 4$)*

Consider a two-step iteration for $z_i^{(k)}$, $1 \le i \le d$:

$$\begin{aligned} z_i^{(k+2)} &= z_i^{(k+1)} \frac{(\sum_{j=1}^d \lambda_j z_j^{(k+1)} - \lambda_i)^2}{D_{k+1}} \\ &= z_i^{(k)} \frac{(\mu_1 - \lambda_i)^2}{D_k} \frac{1}{D_{k+1}} \left(\frac{\mu_1^3 - 2\mu_1\mu_2 + \mu_3}{D_k} - \lambda_i \right)^2, \end{aligned} \tag{7.19}$$

and the corresponding transformation $\psi^2(\cdot)$ defined by $\pi_{k+2} = \psi^2(\pi_k)$. Since $\sum_{i=1}^d z_i^{(k)} = 1$ for all k, we substitute $1 - \sum_{i=1}^{d-1} z_i^{(k)}$ for $z_d^{(k)}$ in (7.19). This defines an operator $\Phi(\cdot) : \mathcal{S}_{d-1} \mapsto \mathcal{S}_{d-1}$ on the $(d-1)$-dimensional canonical simplex

$$\mathcal{S}_{d-1} = \{z = (z_1, \ldots, z_{d-1}) \; / \; z_i \ge 0, \sum_{i=1}^{d-1} z_i \le 1\},$$

which maps $(z_1^{(k)}, \ldots, z_{d-1}^{(k)})$ to $(z_1^{(k+2)}, \ldots, z_{d-1}^{(k+2)})$. Studying the properties of $\psi^2(\cdot)$ is equivalent to studying the properties of $\Phi(\cdot)$.

Theorem 7.2
(i) Stability. *All points in the interval* $\mathcal{I}_S$ *defined by*

$$\mathcal{I}_S =]\frac{1}{2} - s(\lambda_{i^*}),\ \frac{1}{2} + s(\lambda_{i^*})[\,,$$

where

$$s(\lambda) = \frac{\sqrt{(\lambda_d - \lambda)^2 + (\lambda_1 - \lambda)^2}}{2(\lambda_d - \lambda_1)}\,,$$

and i^* *is such that* $|\lambda_{i^*} - \frac{\lambda_1 + \lambda_d}{2}|$ *is minimum over all* λ_i*'s,* $i = 2, \ldots, d-1$*, are stable.*
(ii) Instability. *All points in the set* $\mathcal{I}_U$ *defined by*

$$\mathcal{I}_U = [0,\ \frac{1}{2} - s(\lambda_{i^*})[\ \cup\ ]\frac{1}{2} + s(\lambda_{i^*}),\ 1]$$

are unstable.

Proof.
(i) *Stability.* Take $p \in \mathcal{I}_S$ and consider the fixed point $\boldsymbol{z}(p) = (p, 0, \ldots, 0)$ for $\Phi(\cdot)$, $\boldsymbol{z}(p) \in \mathcal{S}_{d-1}$. Assume that $z_i^{(k)} \leq \beta_1$, $i = 2, \ldots, d-1$ and $|z_1^{(k)} - p| \leq \beta_1$; that is, $\|\boldsymbol{z}^{(k)} - \boldsymbol{z}(p)\|_\infty \leq \beta_1$. Then

$$z_i^{(k+2)} = z_i^{(k)} f(\lambda_i, p)[1 + O(\beta_1)]\,,\ \beta_1 \to 0\,, \tag{7.20}$$

where

$$f(\lambda, p) = \frac{(\mu_1^* - \lambda)^2[(\mu_1^*)^3 - 2\mu_1^*\mu_2^* + \mu_3^* - D(p)\lambda]^2}{[D(p)]^4}$$

with

$$\begin{aligned} \mu_m^* &= \lambda_1^m p + \lambda_d^m (1-p)\,,\ m = 1, 2, 3\,, \\ D(p) &= p(1-p)(\lambda_d - \lambda_1)^2\,. \end{aligned} \tag{7.21}$$

Then, $p \in \mathcal{I}_S$ implies $f(\lambda_i, p) < 1$, $i = 2, \ldots, d-1$. Define $f^*(p) = \max_{i \in \{2, \ldots, d-1\}} f(\lambda_i, p)$ and let K be any constant such that $f^*(p) < K < 1$. Then, from (7.20)

$$\exists \beta_0 \mid \forall \beta < \beta_0\,,\ z_i^{(k+2)} \leq K z_i^{(k)}\,,\ \forall i = 2, \ldots, d-1\,,$$

and thus

$$\forall \beta < \beta_0\,,\ z_i^{(k+2m)} \leq K^m z_i^{(k)}\,,\ \forall i = 2, \ldots, d-1\,,\ \forall m \geq 1\,. \tag{7.22}$$

Now, (7.13) implies $\det[\boldsymbol{M}_3(\pi_k)] \leq C\beta_1$, with C some positive constant. Therefore,

$$|D_{k+1} - D_k| \leq \frac{C\beta_1}{D_k^2}\,. \tag{7.23}$$

Similarly, $||\boldsymbol{z}^{(k+1)} - \boldsymbol{z}(1-p)|| \leq \beta_2$ implies

$$|D_{k+2} - D_{k+1}| \leq \frac{C\beta_2}{D_{k+1}^2} \leq \frac{C\beta_2}{D_k^2}. \tag{7.24}$$

Since $||\boldsymbol{z}^{(k)} - \boldsymbol{z}(p)||_\infty \leq \beta \Rightarrow ||\boldsymbol{z}^{(k+1)} - \boldsymbol{z}(1-p)||_\infty \leq H\beta$, with H some positive constant, (7.23,7.24) imply

$$|D_{k+1} - D_k| \leq C'\beta\,,\ |D_{k+2} - D_{k+1}| \leq C'\beta$$

when $||\boldsymbol{z}^{(k)} - \boldsymbol{z}(p)||_\infty \leq \beta$, with $C' = (C/D_k^2)\max\{1, H\}$. This gives

$$|D_{k+2} - D_k| \leq 2C'\beta\,.$$

Choosing $\beta < \beta_0$, one can then obtain from (7.22)

$$|D_{k+2m+2} - D_{k+2m}| \leq 2C'K^m\beta\,.$$

Therefore, $\forall l > 0$,

$$|D_{k+2l} - D_k| \leq \frac{2C'\beta}{1-K} = O(\beta)\,,\ \beta \to 0\,.$$

Moreover, $||\boldsymbol{z}^{(k)} - \boldsymbol{z}(p)||_\infty \leq \beta \Rightarrow |D_k - D(p)| \leq A\beta$, with A a positive constant and $D(p)$ given by (7.21). Therefore, $\forall l > 0$,

$$|D_{k+2l} - D(p)| \leq \left(\frac{2C'}{1-K} + A\right)\beta\,.$$

This, together with (7.22) implies

$$\forall l > 0\,,\ |z_1^{(k+2l)} - p| < L\beta\,,\ L > 1\,. \tag{7.25}$$

Finally, for any $\epsilon > 0$, define $\alpha = \min\{\epsilon/L, \beta_0/L\}$; (7.22) and (7.25) then imply the stability of the fixed point $\boldsymbol{z}(p)$.
(ii) *Instability.* The Jacobian matrix $\boldsymbol{J}_\Phi$ of $\Phi(\cdot)$ at points $\boldsymbol{z}(p) = (p, 0, \ldots, 0)$ can be computed analytically, and is given by

$$[\boldsymbol{J}_\Phi]_{ij} = \frac{\partial\Phi_i(z)}{\partial z_j}_{|\boldsymbol{z}=\boldsymbol{z}(p)} =$$

$$\begin{cases} 1 & \text{if } i = j = 1 \\ \frac{(\lambda_j-\lambda_1)(\lambda_d-\lambda_j)^2[2\lambda_d(1-p)+2\lambda_1 p-\lambda_1-\lambda_j]}{p(1-p)^2(\lambda_d-\lambda_1)^4} & \text{if } j > 1 \text{ and } i = 1 \\ \frac{[\lambda_d(1-p)+\lambda_1 p-\lambda_j]^2[\lambda_d p+\lambda_1(1-p)-\lambda_j]^2}{p^2(1-p)^2(\lambda_d-\lambda_1)^4} & \text{if } i = j \\ 0 & \text{otherwise}\,. \end{cases}$$

One can easily check that for $p \in \mathcal{I}_U$, $[\boldsymbol{J}_\Phi]_{ii} > 1$ for at least one i in $\{2, \ldots, d-1\}$. Since the $[\boldsymbol{J}_\Phi]_{ii}$'s are eigenvalues of $\boldsymbol{J}_\Phi$, Theorem 15.16 in Hale and Koçak (1991) indicates that $\boldsymbol{z}(p)$ is unstable. ■

Note that $\forall \lambda_1, \ldots, \lambda_d$ the stability interval $\mathcal{I}_S$ contains the interval

$$]\frac{1}{2} - \frac{1}{2\sqrt{2}}, \frac{1}{2} + \frac{1}{2\sqrt{2}}[\approx]0.14645, 0.85355[.$$

When $d = 3$, numerical simulations show that for any initial density of $\boldsymbol{x}^{(1)}$ in $\mathbb{R}^d$ associated with a density of $\boldsymbol{z}^{(1)}$ on $\mathcal{S}_{d-1}$ reasonably spread, the density of attractors $\boldsymbol{z}(p)$ characterized by p can be approximated by the density

$$\phi(p) = C \log \min\{1, [\boldsymbol{J}_\Phi]_{22}\} = \begin{cases} C \log[\boldsymbol{J}_\Phi]_{22} & \text{if } p \in \mathcal{I}_A \\ 0 & \text{otherwise}, \end{cases}$$

where C is a normalisation constant. Figure 7.3 shows the empirical density of attractors (full line) together with $\phi(p)$ (dashed line) in the case $\lambda_1 = 1$, $\lambda_2 = 4$, $\lambda_3 = 10$. The support of this density coincides with the stability interval $\mathcal{I}_S$.

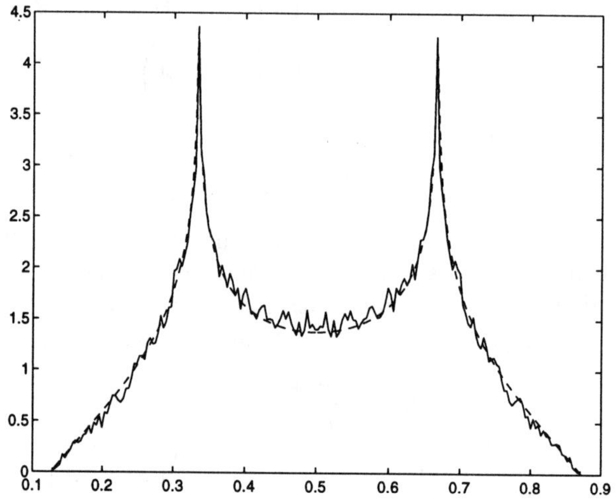

Figure 7.3 *Empirical density of attractors (full line) and* $\phi(p)$ *(dashed line)* $(d = 3, \lambda_1 = 1, \lambda_2 = 4, \lambda_3 = 10)$

When $d > 3$, the density of attractors depends on the initial density of $\boldsymbol{x}^{(1)}$.

7.3 Rate of convergence

Define the convergence rate at iteration k by

$$r_k = \frac{f(\boldsymbol{x}^{(k+1)})}{f(\boldsymbol{x}^{(k)})}, \tag{7.26}$$

where $f(\boldsymbol{x})$ is given by (7.5). Without any loss of generality, one can take $f(\boldsymbol{x}) = \sum_{i=1}^{d} \lambda_i x_i^2$. Rewriting r_k in terms of $\boldsymbol{z}^{(k)}$ defined by (7.10), one gets

$$r_k = 1 - \frac{1}{\mu_1 \mu_{-1}},$$

where μ_m is defined by (7.12). Define the asymptotic rate as

$$R = R(\boldsymbol{x}^{(1)}, \boldsymbol{x}^*) = \lim_{k\to\infty} \left(\prod_{j=1}^{k} r_j \right)^{1/k}. \tag{7.27}$$

Generally, R depends on the initial point $\boldsymbol{x}^{(1)}$ and the optimal point $\boldsymbol{x}^*$. Theorem 7.1 implies that for any fixed $\boldsymbol{x}^*$ and almost all $\boldsymbol{x}^{(1)}$ the asymptotic rate depends only on the attractor (7.18) and is given by

$$R(p) = \left(\frac{f(\boldsymbol{x}^{(k+2)})}{f(\boldsymbol{x}^{(k)})} \right)^{1/2}$$

where $\boldsymbol{x}^{(k)}$ corresponds to π^1 or π^2; see (7.18). This gives

$$R(p) = \frac{p(1-p)(\rho-1)^2}{[p+\rho(1-p)][(1-p)+\rho p]},$$

with $\rho = \lambda_d/\lambda_1$ the condition number of the matrix $\boldsymbol{A}$. The function $R(p)$ is symmetric with respect to $1/2$ and monotonously increasing from 0 to $1/2$. The worst asymptotic rate is thus obtained at $p = 1/2$:

$$R_{\max} = \left(\frac{\rho-1}{\rho+1} \right)^2;$$

see (7.4). Note that from the Kantorovich inequality, see Luenberger (1973), p. 151, $\mu_1\mu_{-1} \le (1+\rho)^2/(4\rho)$, and therefore $\forall \boldsymbol{x}^{(k)}$, $r_k \le R_{\max}$; see (7.4). The worst rate is thus achieved only when $x_1^{(k)} = \pm\rho x_d^{(k)}$, $x_2^{(k)} = \cdots = x_{d-1}^{(k)} = 0$.

Consider now another convergence rate, defined by

$$R' = \lim_{k\to\infty} \left(\prod_{j=1}^{k} r'_j\right)^{1/k},$$

where

$$r'_k = \frac{||\boldsymbol{x}^{(k+1)}||^2}{||\boldsymbol{x}^{(k)}||^2}.$$

Rewriting r'_k in terms of $\boldsymbol{z}^{(k)}$, one gets

$$r'_k = 1 - \frac{2\mu_{-1}}{\mu_1\mu_{-2}} + \frac{1}{\mu_1^2\mu_{-2}}.$$

One can easily check that for almost all $\boldsymbol{x}^{(1)}$ the asymptotic rate R' is equal to $R(p)$, where p defines the attractor.

7.4 Steepest descent with relaxation

The introduction of a relaxation coefficient γ, with $0 < \gamma < 1$, in the steepest-descent algorithm totally changes its behaviour. The algorithm (7.7) then becomes

$$\boldsymbol{x}^{(k+1)} = \boldsymbol{x}^{(k)} - \gamma \frac{(\boldsymbol{g}^{(k)})^T \boldsymbol{g}^{(k)}}{(\boldsymbol{g}^{(k)})^T \boldsymbol{A}\boldsymbol{g}^{(k)}} \boldsymbol{g}^{(k)}.$$

For fixed $\boldsymbol{A}$, depending on the value of γ, the renormalised process either attracts to periodic orbits (the same for almost all starting points) or exhibits a chaotic behaviour. Figure 7.4 (respectively 7.5) presents the classical period-doubling phenomenon in the case $d = 2$ when $\lambda_1 = 1$ and $\lambda_2 = 4$ (respectively $\lambda_2 = 10$). Figure 7.6 (respectively 7.7) gives the asymptotic rate (7.27) as a function of γ in the same situation.

We get now instead of (7.26):

$$r_k(\gamma) = 1 - \frac{\gamma(2-\gamma)}{\mu_1\mu_{-1}}.$$

Note that from the Kantorovich inequality the worst value of the rate is

$$1 - \gamma(2-\gamma)\frac{4\rho}{(1+\rho)^2} > R_{\max},$$

if $\gamma < 1$. However, numerical results show that for γ large enough the asymptotic rate is significantly better than $R_{\max}$. A detailed analysis of the 2-dimensional case gives the following results.

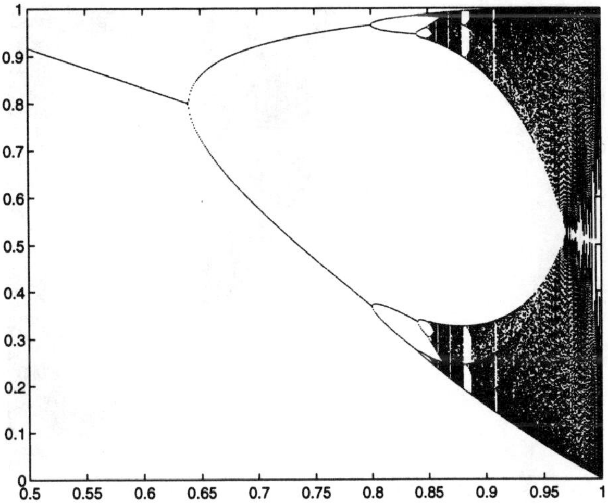

Figure 7.4 *Attractors for* z_1 *as a function of* γ $(d = 2, \lambda_1 = 1, \lambda_2 = 4)$

(i) If $0 < \gamma \le \frac{2}{\rho+1}$, the process attracts to the fixed point $p = 1$ and $R = R(\gamma) = 1 - \frac{2\rho\gamma}{\rho+1}$.

(ii) If $\frac{2}{\rho+1} < \gamma \le \frac{4\rho}{(\rho+1)^2}$, the process attracts to the fixed point $p = \frac{2\rho-\gamma(\rho+1)}{2(\rho-1)}$, and $R(\gamma) = R_{\max}$.

(iii) If $\le \frac{4\rho}{(\rho+1)^2} < \gamma \le \frac{2(\sqrt{2}+1)\rho}{(\rho+1)^2}$, the process attracts to the 2-point cycle (p_1, p_2), with

$$p_{1,2} = \frac{2\rho - \gamma(\rho+1) \pm \sqrt{\gamma(\gamma\rho^2 - 4\rho + 2\gamma\rho + \gamma)}}{2(\rho-1)},$$

and $R(\gamma) = 1 - \gamma$.

(iv) For larger values of γ one observes a classical period-doubling phenomenon; see Figures 7.4 and 7.5.

(v) If $\rho > 3 + 2\sqrt{2} \approx 5.828427$, the process attracts again to a 2-point cycle for values of γ larger than $\gamma_\rho = \frac{8\rho}{(\rho+1)^2}$; see Figure 7.5. For the limiting case $\gamma = \gamma_\rho$, the cycle is given by (p_1', p_2'), with

$$p_{1,2}' = \frac{\rho\left(\rho^2 - 2\rho + 5 \pm 2\sqrt{(\rho^2 - 2\rho + 5)(5\rho^2 - 2\rho + 1)}\right)}{(\rho-1)(\rho+1)^3},$$

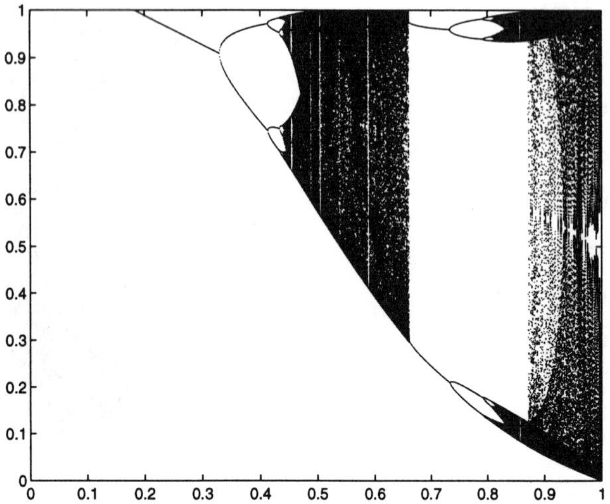

Figure 7.5 *Attractors for* z_1 *as a function of* γ $(d = 2, \lambda_1 = 1, \lambda_2 = 10)$

and the associated asymptotic rate is

$$R(\gamma_\rho) = \frac{\rho^2 - 6\rho + 1}{\rho^2 - 1}.$$

In higher dimensions, repeated numerical trials show that the process typically no longer attracts to the 2-dimensional plane spanned by $(\boldsymbol{u}_1, \boldsymbol{u}_d)$. Figure 7.8 presents the projection of the attractor of $\boldsymbol{z}^{(k)}$ on the plane (z_1, z_3) in the case $d = 3$, $\lambda_1 = 1$, $\lambda_2 = 2$, $\lambda_3 = 4$ and $\gamma = 0.97$. Such a picture is typical for $d > 2$.

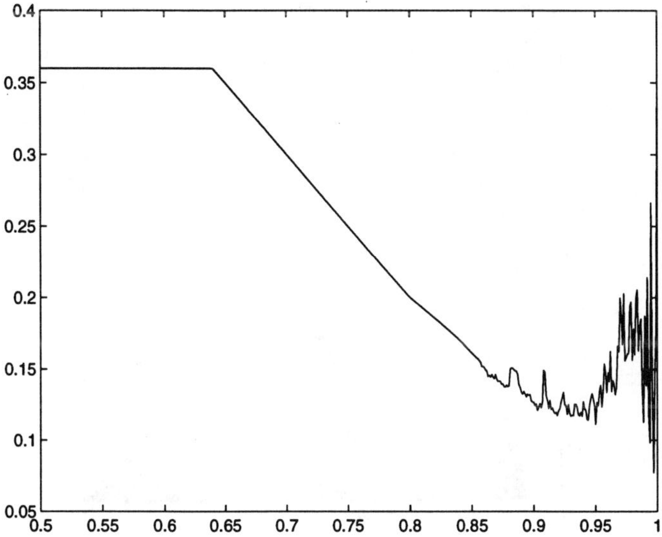

Figure 7.6 *Asymptotic rate (7.27) as a function of* γ *(*$d = 2, \lambda_1 = 1, \lambda_2 = 4$*)*

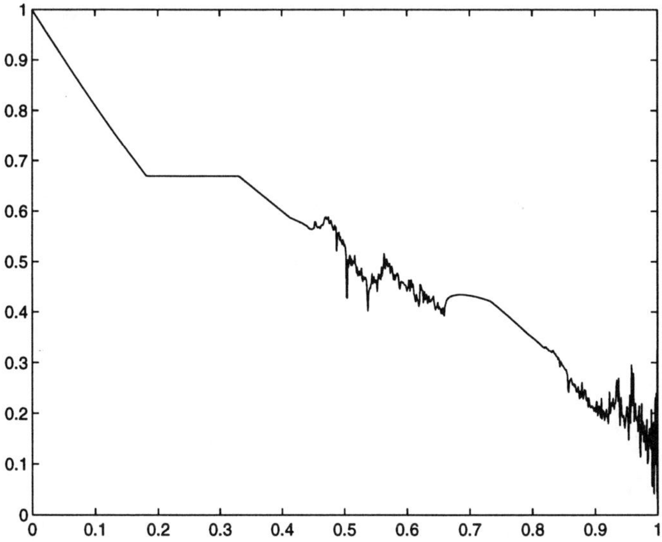

Figure 7.7 *Asymptotic rate (7.27) as a function of* γ *(*$d = 2, \lambda_1 = 1, \lambda_2 = 10$*)*

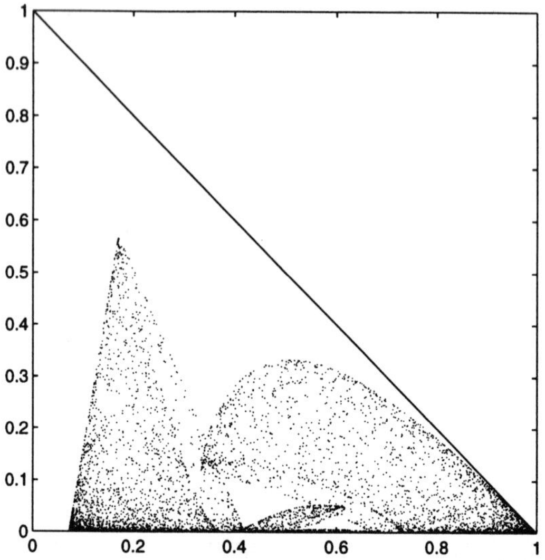

Figure 7.8 *Projection of the attractor on the plane* (z_1, z_3) *(*$d = 3, \lambda_1 = 1, \lambda_2 = 2, \lambda_3 = 4, \gamma = 0.97$*)*

CHAPTER 8

Appendix

8.1 Entropies

8.1.1 Entropies of discrete distributions

Let P be a discrete distribution on a finite or countable set

$$P=\{p_1,p_2,\ldots\}: \ p_i\geq 0,\ i=1,\ldots,N,\ \sum_{i=1}^{N} p_i=1,\ N\leq\infty .$$

The Rényi entropy of order $\gamma \neq 1$ of the distribution P is defined as

$$H_\gamma(P) = \frac{1}{1-\gamma}\log\sum_i p_i^\gamma, \quad -\infty<\gamma<\infty, \quad \gamma\neq 1 . \tag{8.1}$$

An important particular case of (8.1) is the second-order Rényi entropy

$$H_2(P) = -\log\sum_i p_i^2 .$$

If $N < \infty$ and $p_i > 0$ for all $i = 1,\ldots,N$, then the zero-order Rényi entropy is $H_0(P) = \log N$. Shannon entropy can be defined as the limiting case of Rényi entropies of order γ when $\gamma \to 1$:

$$H_1(P) = -\sum_i p_i \log p_i = \lim_{\gamma\to 1} H_\gamma(P) . \tag{8.2}$$

8.1.2 Some properties

Let us give several properties of Rényi and Shannon entropies. We refer to Rényi (1961, 1965) for precise formulations, proofs and related discussions.

(i) For any discrete distribution P and $\gamma \geq 0$

$$0 \leq H_\gamma(P) \leq \log N ,$$

and $H_\gamma(P)=0$ if and only if P is concentrated at a single point;

(ii) if $\gamma > 0$ and $N < \infty$ then

$$H_\gamma(P)=\log N \iff P=\left\{\frac{1}{N},\ldots,\frac{1}{N}\right\};$$

(iii) $H_\gamma(P)$ is a decreasing function of γ;

(iv) for any two discrete distributions P and Q

$$H_\gamma(P \times Q) = H_\gamma(P) + H_\gamma(Q).$$

Under some regularity conditions (Rényi, 1965), the family of entropies $H_\gamma(\cdot)$ provides all possible functionals on the set of discrete distributions which satisfy properties (i), (ii) and (iv), and thus provide all natural support-independent measures of uncertainty for discrete distributions.

8.1.3 Specific properties of the Shannon entropy

The following two important properties hold only when $\gamma = 1$, that is, for Shannon entropy. To formulate them, introduce the bivariate discrete distribution $R=\{r_{ij}\}$ with marginals $P=\{p_i=\sum_j r_{ij}\}$ and $Q=\{q_j=\sum_i r_{ij}\}$.

(v) Subadditivity of Shannon entropy: $H_1(R) \leq H_1(P)+H_1(Q)$;

(vi) Conditioning for Shannon entropy: $H_1(R) = H_1(P)+H_1(Q|P)$ where $H_1(Q|P) = -\sum_{i,j} r_{ij} \log r_{ij}/q_j$.

8.1.4 Entropies of partitions

Let $(\mathcal{X},\mathcal{B},\mu)$ be a probability space with probability measure μ, and $\mathcal{P}=\mathcal{P}_N$ be a partition of $\mathcal{X}$ into $N \leq \infty$ subsets

$$\mathcal{P}_N : \mathcal{X} = \bigcup_{i=1}^{N} \mathcal{X}_i\,, \quad \mu(\mathcal{X}_i \cap \mathcal{X}_j) = 0 \quad \text{for } i \neq j\,.$$

Then we can define the probability distribution P, associated with the partition $\mathcal{P}$,

$$P=\{p_1,\ldots p_N\}, \quad \text{where } p_i=\mu(\mathcal{X}_i) \geq 0,\ \sum_i p_i = 1,$$

and therefore the entropies

$$H_\gamma(\mathcal{P}) = H_\gamma(\mathcal{P},\mu) = \begin{cases} \frac{1}{1-\gamma}\log\sum_i p_i^\gamma & \text{if } \gamma \neq 1 \\ -\sum_i p_i \log p_i & \text{if } \gamma = 1\,. \end{cases} \tag{8.3}$$

In several parts of the book we study the asymptotic behaviour of Rényi entropies of sequences of partitions generated by different search algorithms. In some particular cases this study can be related to that of the Mellin transform of a sequence of measures related to $\mathcal{P}_N$. More precisely, let ν_N be the measure assigning the weight 1 to p_i, $i = 1,\ldots,N$. Then the Mellin transform of ν_N is

$$M(\nu_N)(\gamma) = \sum_i p_i^\gamma\,,$$

and therefore the study of the asymptotic properties of $H_\gamma(\mathcal{P}_N)$ for all $\gamma > 0$, when $N \to \infty$, is equivalent to the study of the Mellin transform of the sequence ν_N.

8.1.5 Three one-dimensional examples

Let $\mathcal{X} = [0,1)$, μ be the Lebesgue measure $\mu_{\mathcal{L}}$, $N < \infty$,

$$\mathcal{X} = \bigcup_{i=1}^{N} I_i, \quad I_i = [x_{i-1}, x_i)\,,$$

where $0 = x_0 \leq x_1 \leq \cdots \leq x_N = 1$, $p_i = |I_i| = x_i - x_{i-1}$.

1. Let $x_j = j/N$ for all $j = 0,\ldots,N$. Then $p_i = 1/N$ and therefore $H_\gamma(\mathcal{P}_N) = \log N$ for all real γ. According to property (i) of Section 8.1.2, this value of $H_\gamma(\mathcal{P}_N)$ is maximal for any $\gamma \geq 0$; we thus can consider this partition as the most uniform.

2. Let x_i ($i = 1,\ldots,N-1$) be the order statistics corresponding to independent uniformly distributed random points in $[0,1)$. Then the results of Drobot (1981) imply that for any $\gamma > 1$,

$$\log N - H_\gamma(\mathcal{P}_N) \to \frac{\log\Gamma(\gamma+1)}{\gamma - 1} \quad \text{a.s. when} \quad N \to \infty\,, \tag{8.4}$$

where $\Gamma(\cdot)$ is the gamma function. Roughly speaking, this means that the partition of $[0,1)$, generated by uniformly distributed random points, is very close to the uniform partition of Example 1 above, if the uniformity of partitions is measured via the entropies

$H_\gamma(\mathcal{P}_N)$. For Shannon entropy, (8.2) and (8.4) imply

$$\log N - H_1(\mathcal{P}_N) \to 1 - \gamma_0 \quad \text{a.s.} \quad \text{when} \quad N \to \infty,$$

where $\gamma_0 \simeq 0.5772$ is Euler's constant.

3. Let x_i be ordered rationals p/q with p and q mutually prime, $p \le q$, $q \le n$ (Farey sequence of order n). Then

$$N = N(n) = \sum_{k=1}^{n} \varphi(k) \simeq \frac{3}{\pi^2} n^2, \quad n \to \infty,$$

where $\varphi(\cdot)$ is the Euler function, and, according to Kargaev and Zhigljavsky (1997),

$$H_\gamma(\mathcal{P}_N) = \begin{cases} \log N + O(1) & \text{if } \gamma < 2 \\ \log N - \log\log N + O(1) & \text{if } \gamma = 2 \\ \frac{\gamma}{2(\gamma-1)} \log N + O(1) & \text{if } \gamma > 2\,. \end{cases}$$

Comparison with (8.4) leads us to the conclusion that the partition of $[0, 1)$ generated by the rationals p/q with $q \le n$ is much less uniform, if uniformity is measured via the entropies $H_\gamma(\mathcal{P}_N)$, than the partition generated by the same number of independent uniformly distributed random points. Note also that Rényi entropies of order $\gamma < 2$, including Shannon entropy, are not sensitive enough to detect the lack of uniformity of Farey sequences.

8.2 Ergodic theory

In this section, we consider basic properties of dynamical systems. For a more detailed exposition, we refer to Petersen (1983), Mañé (1987), Ruelle (1989), Bedford, Keane and Series (1991), Ott (1993) and Aaronson (1997).

8.2.1 Dynamical systems

Let $(\mathcal{X}, \mathcal{B}, \mu)$ be a measure space. Consider a measurable transformation $T(\cdot)$ of $(\mathcal{X}, \mathcal{B}, \mu)$ to itself, and define the dynamical system

$$\boldsymbol{x}_{n+1} = T(\boldsymbol{x}_n), \quad n = 1, 2, \ldots$$

The mapping $T(\cdot)$ is called *measure preserving* if

$$\mu(\mathcal{A}) = \mu(T^{-1}(\mathcal{A}))$$

for any $\mathcal{A} \in \mathcal{B}$. The measure μ is then called *invariant* with respect to $T(\cdot)$. The mapping $T(\cdot)$ and the dynamical system are called *ergodic* with respect to μ if $T^{-1}(\mathcal{A}) = \mathcal{A}$ for some measurable set $\mathcal{A} \subseteq \mathcal{X} \Rightarrow$ either $\mu(\mathcal{A}) = 0$ or $\mu(\mathcal{X} \setminus \mathcal{A}) = 0$. The invariant measure is then called *T-ergodic*, and will be denoted μ_T. We refer to Chapter 1 of Aaronson for a detailed study of existence and uniqueness of invariant measures. See also Section 8.2.4.

If $\mu(\mathcal{X}) < \infty$, then the measure can be normalised and considered as a probability measure. The measure μ is then ergodic if it cannot be written as $\frac{1}{2}\mu_1 + \frac{1}{2}\mu_2$, where μ_1 and μ_2 are again invariant probability measures and $\mu_1 \neq \mu_2$. Suppose an invariant probability measure μ_T has a density ϕ with respect to the Lebesgue measure. Then one can write the following equation, which expresses invariance with respect to the movement of the process forward in time:

$$\phi(\boldsymbol{x}) = \int \phi(\boldsymbol{y})\delta[\boldsymbol{x} - T(\boldsymbol{y})]d\boldsymbol{y},$$

where $\delta(\cdot)$ is the Dirac delta function. This equation is called the *Frobenius-Perron equation*. In many cases the following differential form of the Frobenius-Perron equation is more useful:

$$\phi(\boldsymbol{x}) = \sum_k \frac{\phi(\boldsymbol{y}_k)}{|\det \boldsymbol{J}_T(\boldsymbol{y}_k)|}, \tag{8.5}$$

where $\{\boldsymbol{y}_k\} = T^{-1}(\boldsymbol{x})$ and where $\boldsymbol{J}_T$ is the Jacobian matrix of T with entries

$$[\boldsymbol{J}_T(\boldsymbol{x})]_{ij} = \frac{\partial[T(\boldsymbol{x})]_i}{\partial[\boldsymbol{x}]_j}.$$

The *transfer operator* of order γ of the dynamical system T is defined by

$$G_\gamma[f](\boldsymbol{x}) = \sum_k \frac{f(\boldsymbol{y}_k)}{|\det \boldsymbol{J}_T(\boldsymbol{y}_k)|^\gamma}.$$

When $\gamma = 1$, $G_1[\cdot]$ is called the *Frobenius-Perron operator*, and the invariant density ϕ is the eigenfunction associated with the maximum eigenvalue 1. Conditions for the maximum eigenvalue of $G_\gamma[\cdot]$ being simple are given by the *Krasnoselskii Theorem*; Krasnoselskii (1964), Mayer (1991). These conditions imply in particular the uniqueness of the invariant density ϕ.

Let $\mathcal{F}$ be a Banach space of real-valued functions, $\mathcal{K}$ be a subset of positive functions and $G : \mathcal{F} \rightarrow \mathcal{F}$ be a compact u_0-positive operator, with $u_0 \in \mathcal{K}$; that is, there exist α and β positive and p

integer such than

$$\text{for any nonnegative } f \neq 0\,, \; \alpha u_0 \leq G^p[f] \leq \beta u_0\,. \qquad (8.6)$$

Then, the Krasnoselskii Theorem states that there exists exactly one eigenfunction $f_1 \in \mathcal{K}$ and λ_1 real positive such that $G[f_1] = \lambda_1 f_1$; the eigenvalue λ_1 is simple and in absolute value larger than the modulus of all other eigenvalues of G. Also, for any $f \in \mathcal{F}$, one has

$$\lim_{n\to\infty} \frac{G^n[f]}{\lambda_1^n} = L(f) f_1\,,$$

where $G^* L = \lambda_1 L$, with G^* the conjugate operator of G. The *Frobenius-Perron Theorem* corresponds to the particular case when G is a square matrix with nonnegative elements. Then, u_0 is the vector **1** and the condition (8.6) corresponds to the *strong mixing* condition; see Section 4.5.4. If the strong mixing condition does not hold but the matrix is irreducible, then we only have that the maximum eigenvalue λ_1 is simple. Its associated eigenvector has positive elements, but the modulus of the other eigenvalues are not strictly smaller than λ_1; see Gantmacher (1966), vol. 2, p. 49.

Birkhoff's ergodic Theorem, see Keane (1991), states that if the dynamical system is ergodic, with μ_T a finite ergodic invariant measure, and $f \in L^1(\mathcal{X}, \mu_T)$, then

$$\lim_{N\to\infty} \frac{1}{N} \sum_{n=1}^{N} f(\boldsymbol{x}_n) = \int_{\mathcal{X}} f(\boldsymbol{x}) \mu_T(d\boldsymbol{x})$$

for μ_T-almost all initial points $\boldsymbol{x}_1 \in \mathcal{X}$. A detailed description of several important ergodic theorems can be found in Chapter 2 of Aaronson (1997).

A dynamical system is usually called *chaotic* if it is ergodic and two orbits $(\boldsymbol{x}_n)$ and $(\boldsymbol{x}'_n)$ starting at close points $\boldsymbol{x}_1 \simeq \boldsymbol{x}'_1$ exponentially diverge. A classical measure of the rate of the divergence is given by the so-called Lyapunov exponents.

8.2.2 Lyapunov exponents

Let $T(\cdot)$ be an ergodic one-dimensional map, with μ_T its invariant measure. If $T(\cdot)$ is μ_T-almost everywhere differentiable, with $T'(\cdot)$ its derivative, and $\int \log |T'(x)| \mu_T(dx) < \infty$, then the limit

$$\Lambda = \lim_{n\to\infty} \frac{1}{n} \sum_{i=1}^{n} \log |T'(\boldsymbol{x}_i)|$$

exists for μ_T almost initial points x_1. This limit is called the Lyapunov exponent of $T(\cdot)$, and is equal to

$$\Lambda = \int \log |T'(x)| \mu_T(dx) .$$

The Lyapunov exponents for a d-dimensional map $T : \mathcal{X} \to \mathcal{X}$ μ_T-almost everywhere differentiable can be defined as follows. Let $\boldsymbol{x}_1 \in \mathcal{X}$ be an initial point in $\mathcal{X}$, $(\boldsymbol{x}_i)$ be the corresponding orbit, and $\boldsymbol{u}_1 \in \mathbb{R}^d$, $||\boldsymbol{u}_1|| = 1$, be an arbitrary vector of unit length. If we consider an infinitesimal displacement from $\boldsymbol{x}_1$ in the direction of a tangent vector $\boldsymbol{u}_1$, then the evolution of the tangent vector, given by $\boldsymbol{u}_{n+1} = \boldsymbol{J}_T(\boldsymbol{x}_n)\boldsymbol{u}_n$, determines the evolution of the infinitesimal displacement of the orbit from the unperturbed point $\boldsymbol{x}_n$. This gives $\boldsymbol{u}_{n+1} = \boldsymbol{J}_{T^n}(\boldsymbol{x}_1)\boldsymbol{u}_1$, where

$$\boldsymbol{J}_{T^n}(\boldsymbol{x}_1) = \boldsymbol{J}_T(\boldsymbol{x}_n)\, \boldsymbol{J}_T(\boldsymbol{x}_{n-1}) \ldots \boldsymbol{J}_T(\boldsymbol{x}_1).$$

The Lyapunov exponent for initial condition $\boldsymbol{x}_1$ and initial orientation $\boldsymbol{u}_1$, $||\boldsymbol{u}_1|| = 1$, is defined as

$$\Lambda(\boldsymbol{x}_1, \boldsymbol{u}_1) = \lim_{n\to\infty} \frac{1}{n} \log\left(\frac{||\boldsymbol{u}_{n+1}||}{||\boldsymbol{u}_1||}\right) = \lim_{n\to\infty} \frac{1}{n} \log(||\boldsymbol{J}_{T^n}(\boldsymbol{x}_1)\, \boldsymbol{u}_1||). \tag{8.7}$$

There are d or less different Lyapunov exponents for a given $\boldsymbol{x}_1$, and which one of them applies depends on the initial orientation $\boldsymbol{u}_1$.

If the Jacobian matrices $\boldsymbol{J}_{T^n}(\boldsymbol{x}_1)$ have real eigenvalues for all n and the limits in (8.7) exist, then the d Lyapunov exponents corresponding to the point $\boldsymbol{x}_1$ can be computed as

$$\Lambda_i(\boldsymbol{x}_1) = \lim_{n\to\infty} \frac{1}{n} \log[\lambda_{i,n}(\boldsymbol{x}_1)], \quad ,i = 1, \ldots, d, \tag{8.8}$$

where $\lambda_{1,n}(\boldsymbol{x}_1) \geq \ldots \geq \lambda_{d,n}(\boldsymbol{x}_1)$ are the absolute values of the eigenvalues of $\boldsymbol{J}_{T^n}(\boldsymbol{x}_1)$ taken in decreasing order.

Oseledets' Theorem, see Jensen (1993), guarantees the existence of the limits used in the definition of the Lyapunov exponents under very general conditions. In particular, if $(\boldsymbol{x}_n)$ is an ergodic sequence, with μ_T as invariant measure, then the Lyapunov exponents $\Lambda_i(\boldsymbol{x}_1)$ are the same for μ_T-almost every $\boldsymbol{x}_1$. In this case the Lyapunov exponents can be denoted $\Lambda_1, \ldots, \Lambda_d$, with $\Lambda_1 \geq \ldots \geq \Lambda_d$.

If the largest Lyapunov exponent is positive, then the dynamical system is *chaotic*. Some dynamical systems asymptotically approach fractal attractors. There are several definitions of fractal dimensions. We only quote one of them. Let K be the largest integer such that $\sum_{j=1}^{K} \Lambda_j \geq 0$; then the *Lyapunov dimension* is defined as

$$D_L = K + \frac{1}{|\Lambda_{K+1}|} \sum_{j=1}^{K} \Lambda_j .$$

8.2.3 Entropies of dynamical systems

Let $\mathcal{X}$ be a compact subset of $\mathbb{R}^d$, $\mathcal{B}$ be the σ-algebra of Borel subsets of $\mathcal{X}$, $\mu_{\mathcal{L}}$ be the Lebesgue measure which we shall call the uniform measure on $\mathcal{X}$, with $\mu_{\mathcal{L}}(\mathcal{X}) = 1$, and $T : \mathcal{X} \to \mathcal{X}$ be an ergodic map, with an invariant measure μ_T. Consider a partition $\mathcal{P}$ of $\mathcal{X}$:

$$\mathcal{P} : \mathcal{X} = \bigcup_{i=1}^{N} \mathcal{X}_i .$$

Application of the mapping $T(\cdot)$ k times, starting at $\boldsymbol{x}_0$, gives another partition for $\mathcal{X}$:

$$\mathcal{P}_k = T^{-k}(\mathcal{P}) = \bigcup_{i=1}^{N} \{\boldsymbol{x}_0 \in \mathcal{X} \ / \ \boldsymbol{x}_k = T^k(\boldsymbol{x}_0) \in \mathcal{X}_i\} .$$

The join of partitions $\mathcal{P}, \mathcal{P}_1, \ldots, \mathcal{P}_{n-1}$ is then defined as

$$\begin{aligned} \mathcal{Q}_n(\mathcal{P}) &= \mathcal{P} \vee \mathcal{P}_1 \vee \ldots \vee \mathcal{P}_{n-1} \\ &= \{\cup_{i_1,\ldots,i_n} \mathcal{X}_{i_1,\ldots,i_n}, \ i_k \in \{1, \ldots, m\}, k = 1, \ldots, n\} , \end{aligned} \qquad (8.9)$$

where

$$\mathcal{X}_{i_1,\ldots,i_n} = \{\boldsymbol{x}_0 \in \mathcal{X} \ / \ \boldsymbol{x}_0 \in \mathcal{X}_{i_1}, \ldots, \boldsymbol{x}_{n-1} \in \mathcal{X}_{i_n}\} .$$

Consider the entropies $H_\gamma(\mathcal{P}, \mu)$ defined in (8.3), and for any fixed $\gamma \geq 0$, measure μ and initial partition $\mathcal{P}$ define

$$h_\gamma(T, \mathcal{P}, \mu) = \lim_{n \to \infty} \frac{1}{n} H_\gamma(\mathcal{Q}_n(\mathcal{P}), \mu) \qquad (8.10)$$

if the limit exists. Note that according to classical results in ergodic theory, see Keane (1991), the limit in (8.10) certainly exists in two cases: (i) $\gamma = 0$, $\mu = \mu_{\mathcal{L}}$, and (ii) $\gamma = 1$, $\mu = \mu_T$. In a more general case, the following theorem holds.

Theorem 8.1 *Let* $(\mathcal{X}, \mathcal{B}, \mu)$ *be a measure space,* $\mu(\mathcal{X}) = 1$, $T(\cdot)$ *be an endomorphism on* $(\mathcal{X}, \mathcal{B}, \mu)$, $\gamma \geq 0$ *be a fixed number, and*

$\mathcal{P}$ be a partition of $\mathcal{X}$ such that the limit in the right-hand side of (8.10) exists. Let μ' be another finite measure on $(\mathcal{X}, \mathcal{B})$ such that for some positive constants $0 < c_1 < c_2 < \infty$

$$c_1\mu(A) \le \mu'(A) \le c_2\mu(A)$$

for any $A \in \mathcal{B}$. Then the limit

$$h_\gamma(T, \mathcal{P}, \mu') = \lim_{n\to\infty} \frac{1}{n} H_\gamma(\mathcal{Q}_n(\mathcal{P}), \mu')$$

exists and $h_\gamma(T, \mathcal{P}, \mu') = h_\gamma(T, \mathcal{P}, \mu)$.

Proof. Let $\gamma > 0$, $\gamma \ne 1$. Denote

$$p_{i_1,\dots,i_n} = \mu\left(\mathcal{X}_{i_1,\dots,i_n}\right), \quad p'_{i_1,\dots,i_n} = \mu'\left(\mathcal{X}_{i_1,\dots,i_n}\right).$$

Then

$$H_\gamma(\mathcal{Q}_n(\mathcal{P}), \mu') = \frac{1}{1-\gamma} \log \sum_{i_1,\dots,i_n} (p'_{i_1,\dots,i_n})^\gamma$$

$$\le \frac{1}{1-\gamma} \log c_2^\gamma \sum_{i_1,\dots,i_n} p^\gamma_{i_1,\dots,i_n} = \frac{\gamma}{1-\gamma} \log c_2 + H_\gamma(\mathcal{Q}_n(\mathcal{P}), \mu).$$

Analogously,

$$\frac{\gamma}{1-\gamma} \log c_1 + H_\gamma(\mathcal{Q}_n(\mathcal{P}), \mu) \le H_\gamma(\mathcal{Q}_n(\mathcal{P}), \mu').$$

Therefore,
$h_\gamma(T, \mathcal{P}, \mu) =$

$$\lim_{n\to\infty} \frac{1}{n} H_\gamma(\mathcal{Q}_n(\mathcal{P}), \mu) = \lim_{n\to\infty} \frac{1}{n} \left(\frac{\gamma}{1-\gamma} \log c_1 + H_\gamma(\mathcal{Q}_n(\mathcal{P}), \mu)\right)$$

$$\le \lim_{n\to\infty} \frac{1}{n} H_\gamma(\mathcal{Q}_n(\mathcal{P}), \mu') \le \lim_{n\to\infty} \frac{1}{n} \left(\frac{\gamma}{1-\gamma} \log c_2 + H_\gamma(\mathcal{Q}_n(\mathcal{P}), \mu)\right)$$

$$= \lim_{n\to\infty} \frac{1}{n} H_\gamma(\mathcal{Q}_n(\mathcal{P}), \mu) = h_\gamma(T, \mathcal{P}, \mu),$$

which implies that the limit $\lim_{n\to\infty} \frac{1}{n} H_\gamma(\mathcal{Q}_n(\mathcal{P}), \mu')$ exists and equals $h_\gamma(T, \mathcal{P}, \mu)$.

The cases $\gamma = 0$ and $\gamma = 1$ are treated analogously. ■

The Kolmogorov, or metric, entropy of the dynamical system $(\boldsymbol{x}_{n+1} = T(\boldsymbol{x}_n))$ with respect to μ_T is defined as

$$h_1(T, \mu_T) = \sup_{\mathcal{P}} h_1(T, \mathcal{P}, \mu_T), \tag{8.11}$$

where the supremum is taken over all partitions of $\mathcal{X}$ such that the limit in (8.10) exists. The topological entropy of this system equals

$$h_0(T) = \sup_{\mathcal{P},\mu} h_0(T,\mathcal{P},\mu) = \sup_{\mathcal{P}} h_0(T,\mathcal{P},\mu_{\mathcal{L}})\,. \tag{8.12}$$

Analogously to (8.11) and (8.12), we define the entropy of order $\gamma \geq 0$ of the dynamical system $(\boldsymbol{x}_{n+1} = T(\boldsymbol{x}_n))$ with respect to a measure μ:

$$h_\gamma(T,\mu) = \sup_{\mathcal{P}} h_\gamma(T,\mathcal{P},\mu)\,, \tag{8.13}$$

where the supremum is taken over all partitions of $\mathcal{X}$ such that the limit in (8.10) exists.

A partition $\mathcal{P}$ is called *generating* with respect to $T(\cdot)$ if the σ-algebra generated by $\cup_{n=1}^{\infty}\mathcal{Q}_n(\mathcal{P})$, where $\mathcal{Q}_n(\mathcal{P})$ is given by (8.9), contains all Borel sets in $\mathcal{X}$ up to sets of zero Lebesgue measure. The supremum over $\mathcal{P}$ in (8.13) is achieved at any generating partition. The dynamical systems for which entropies have been studied in previous chapters are such that $T(\cdot)$ is a *Markov map*, see the next section, and the corresponding Markov partition is generating.

8.2.4 Markov maps

Consider a measurable transformation $T(\cdot)$ of the measure space $(\mathcal{X},\mathcal{B},\mu)$. A countable generating partition $\mathcal{P} = \{\mathcal{A}_i\} \subset \mathcal{B}$ is called *basic* for $T(\cdot)$ if the restriction of $T(\cdot)$ to any $\mathcal{A} \in \mathcal{P}$ is an invertible map. If, in addition, for each $\mathcal{A}_i$ in $\mathcal{P}$ there exists a set of indices Ω_i such that

$$T(\mathcal{A}_i) = (\cup_{j\in\Omega_i}\mathcal{A}_j) \cup \mathcal{A}'\,,$$

with $\mathcal{A}'$ such that $\mu(\mathcal{A}') = 0$ and $\mathcal{A}_j \in \mathcal{P}$ for all j in Ω_i, then $\mathcal{P}$ is called a *Markov partition* and $T(\cdot)$ a *Markov map*. A detailed exposition of Markov maps and their properties is given in Chapter 4 of Aaronson (1997). In what follows, we restrict our presentation to one-dimensional maps.

One-dimensional Markov maps

Take $\mathcal{X} = [0,1]$, and consider the measure space $([0,1],\mathcal{B},\mu_{\mathcal{L}})$. Assume that $T(\cdot)$ is a Markov map, with Markov partition $\mathcal{P} = \{\mathcal{A}_i\}$ such that all $\mathcal{A}_i$'s are intervals and the restriction of $T(\cdot)$ on each $\mathcal{A}_i$ admits a twice continuously differentiable extension on the closure of $\mathcal{A}_i$. Also assume that

$$M = \sup_{\mathcal{A}_i} \sup_{(y,z)\in\mathcal{A}_i^2} \frac{|T''(z)|}{|T'(y)|^2} < \infty \,,$$

and

$$\inf_x |(T^n)'(x)| > 1 \tag{8.14}$$

for some n. Then $T(\cdot)$ admits an invariant probability measure μ_T, with density ϕ satisfying $\forall x \in [0,1]$: $0 < \phi(x) < \infty$. This result is known as the "*Folklore Theorem*"; see Bowen (1979) and Adler (1991).

It directly applies, for instance, to the case of piecewise linear mappings of the interval considered in Section 4.5. The partition $\mathcal{S}^\infty = \{\mathcal{I}_i\}$ is a Markov partition, $T''(z) = 0$ and $T'(z) > 1$ for all z in each $\mathcal{I}_i$.

In the case of the Gauss map, see Section 3.2.2, the partition $\mathcal{P} = \{[1/(k+1), 1/k), k = 1, 2, \ldots\}$ is obviously a Markov partition, $M = 16$ and $\inf_x |(T^2)'(x)| \geq 9/4$; see Cornfeld, Fomin and Sinai (1982), Chapter 7.

Maps satisfying the condition (8.14) are called *eventually expanding*. The Farey map of Section 3.2.2 is not eventually expanding, and the theorem above does not directly apply. However, a generalisation of this theorem is given in Bowen (1979), which shows that an invariant measure exists for the Farey map, absolutely continuous with respect to the Lebesgue measure, but not finite.

8.3 Section-invariant numbers

An ordered m-collection of numbers $\mathcal{U} = \{v_1, \ldots, v_m\}$ is called *section invariant* if $0 < v_1 < \cdots < v_m < 1$ and there exists an associated m-collection of numbers $\mathcal{U}' = \{v'_1, \ldots, v'_m\}$ in $(0,1)$ such that for every $j = 1, \ldots, m$

$$R(v_j, v'_j) \in \mathcal{U} \text{ and } L(v_j, v'_j) \in \mathcal{U} \,,$$

where

$$L(v, v') = \frac{\max(v, v') - \min(v, v')}{1 - \min(v, v')} \,, \quad R(v, v') = \frac{\min(v, v')}{\max(v, v')}$$

respectively correspond to Left and Right deletions in a second-order line-search algorithm; see Sections 2.2.3.

Let us present several examples of section-invariant sets in the form of $2 \times |\mathcal{U}|$-tables where the first column contains the elements of $\mathcal{U}$ and the second one contains the elements of $\mathcal{U}'$.

(a)

v	v'
$1-\varphi$	φ
φ	$1-\varphi$

(b)

v	v'
1/2	3/4
2/3	1/3

(c)

v	v'
1/3	2/3
1/2	1/4

(d)

v	v'
$1-\psi$	$1-\psi+\psi^2$
ψ	$\psi-\psi^2$

Table 8.1 *Section-invariant numbers with* $|\mathcal{U}| = 2$

Table 8.1(a) corresponds to the Golden-Section algorithm. Here $\varphi = (\sqrt{5}-1)/2 \simeq 0.61804$ is the smallest root of the polynomial $t^2 + t - 1$ and is called the Golden-Section ratio. In Table 8.1(d) $\psi \simeq 0.5698$ is the minimal positive root of the polynomial $\psi^3 - \psi^2 + 2\psi - 1$.

Table 8.1 gives all possible sets of section-invariant numbers with $|\mathcal{U}| = 2$. The situation for $|\mathcal{U}| > 2$ is more complex because of the range of different configurations available under (L) or (R). The authors have found all sets of section-invariant numbers for $|\mathcal{U}| = 3$ by simple computer enumeration. The general structure is as in Table 8.2.

v	v'	$L(v, v')$	$R(v, v')$
a	a'	$L(a, a')$	$R(a, a')$
b	b'	$L(b, b')$	$R(b, b')$
c	c'	$L(c, c')$	$R(c, c')$

Table 8.2 *Section-invariant numbers with* $|\mathcal{U}| = 3$

Let $S^{(1)} = (a, a')$, $S^{(2)} = (b, b')$, $S^{(3)} = (c, c')$. Any given structure, that is, $L(S^{(j)})$ and $R(S^{(j)})$ set to fixed elements in $\mathcal{U} =$

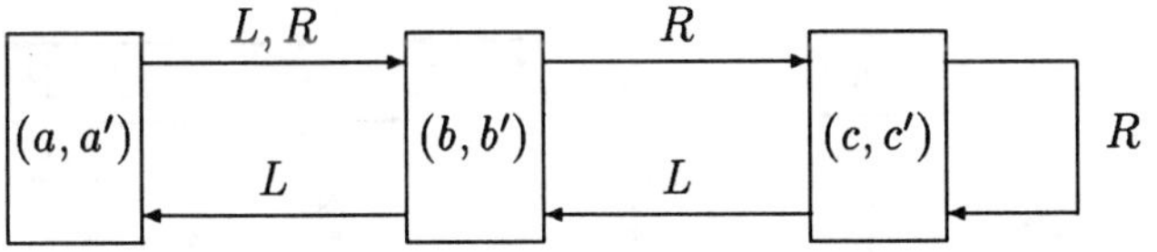

Figure 8.1 *The directed graph in Example 8.1*

$\{a, b, c\}$, $j = 1, 2, 3$, gives six polynomial equations of first or second degree with six unknowns (the elements of $\mathcal{U}$ and $\mathcal{U}'$). Elimination of unknowns leads to one polynomial equation in one variable. This equation is typically of degree one, two or three. Its maximum degree, seven, is obtained when $L(S^{(1)}) = R(S^{(1)}) = b$, $L(S^{(2)}) = R(S^{(2)}) = c$ and $L(S^{(3)}) = R(S^{(3)}) = a$. Since $a \leq b \leq c$, we have $a' > a$ and $c' < c$. Because of symmetry with respect to $1/2$, we restrict ourselves to the case $b' > b$. These conditions imply $R(S^{(1)}) \in \{b, c\}$, $R(S^{(2)}) = c$, $L(S^{(3)}) \in \{a, b\}$, and $L(S^{(1)}), L(S^{(2)}), R(S^{(3)}) \in \{a, b, c\}$, which gives 108 possible cases. Reachability of a (resp. of b) implies that a (resp. b) should appear at least once in $\{L(S^{(2)}), L(S^{(3)}), R(S^{(3)})\}$ (resp. in $\{L(S^{(1)}), R(S^{(1)}), L(S^{(3)}), R(S^{(3)})\}$). These conditions remove respectively 24 and 12 cases. Among the remaining 72 structures, 4 coincide with Table 8.1(a), 12 with Table 8.1(b) and 2 with Table 8.1(d) (which means that for 18 structures $a = b$ or $b = c$). Additionally, 6 structures yield $b < a$, and correspond to other admissible structures by permutation of indices. Finally, one structure has no solution; it corresponds to $L(S^{(1)}) = a, R(S^{(1)}) = b, L(S^{(2)}) = a, R(S^{(2)}) = c, L(S^{(3)}) = b, R(S^{(3)}) = a$. We thus end up with the 47 cases presented in Table 8.3. We also give in this table the value of the log-rate ϱ defined by (4.3). The numerical values of a, b, c, a', b', c' are obtained as roots of various polynomials, and we examine one example to illustrate the construction of this table.

Example 8.1 *The structure is defined by Table 8.4, and the corresponding graph is presented in Figure 8.1.*

The analytical expressions for the components of the table are determined as follows. From the definitions of $L(v_j, v_j')$ and $R(v_j, v_j')$, see (2.12), we have $b = (a'-a)/(1-a)$, $b = a/a'$. Eliminating a' we obtain $a = b^2/(b^2 - b + 1)$. Taking $b' > b$, the convention for Table

a	a'	L_a, R_a	b	b'	L_b, R_b	c	c'	L_c, R_c	ϱ
.245	.570	b , b	.430	.570	a , c	.755	.570	b , c	.557
.317	.534	a , b	.594	.817	c , c	.682	.217	b , a	.317
.317	.783	c , b	.406	.594	a , c	.682	.466	b , c	.483
.363	.637	b , c	.430	.755	c , c	.570	.325	a , c	.421
.366	.698	b , b	.524	.826	c , c	.634	.232	b , a	.321
.382	.618	a , c	.500	.809	c , c	.618	.236	b , a	.365
.382	.764	c , b	.500	.809	c , c	.618	.382	a , c	.393
.382	.764	c , b	.500	.809	c , c	.618	.236	b , a	.423
.382	.618	a , b	.618	.854	b , c	.724	.276	b , a	.336
.382	.618	a , b	.618	.764	a , c	.809	.500	b , b	.481
.393	.723	b , b	.544	.839	c , c	.648	.420	a , c	.395
.400	.800	c , b	.500	.750	b , c	.667	.445	a , c	.410
.406	.682	b , c	.466	.783	c , c	.594	.241	b , a	.348
.414	.657	a , b	.631	.892	c , c	.707	.500	a , c	.428
.409	.737	b , b	.555	.802	b , c	.692	.479	a , c	.420
.414	.828	c , b	.500	.707	a , c	.707	.500	a , c	.420
.423	.667	a , b	.634	.866	b , c	.732	.536	a , c	.438
.430	.755	b , b	.570	.755	a , c	.755	.570	a , c	.451
.430	.755	b , b	.570	.755	a , c	.755	.430	b , b	.423
.430	.755	b , b	.570	.815	b , c	.699	.301	b , a	.366
.445	.692	a , b	.643	.802	a , c	.802	.643	a , c	.479
.447	.724	b , c	.500	.809	c , c	.618	.309	a , b	.385
.453	.829	c , b	.547	.794	b , c	.688	.312	b , a	.382
.456	.704	a , c	.544	.839	c , c	.648	.352	a , b	.414
.459	.848	c , b	.541	.752	a , c	.720	.389	b , b	.390
.460	.817	c , b	.563	.852	c , c	.661	.372	a , b	.351
.466	.783	b , b	.594	.871	c , c	.682	.406	a , b	.347
.475	.724	a , b	.656	.905	c , c	.724	.475	a , b	.362
.476	.751	b , c	.524	.826	c , c	.634	.302	a , a	.378
.500	.854	c , b	.586	.828	b , c	.707	.414	a , b	.349
.500	.809	b , b	.618	.854	b , c	.724	.447	a , b	.353
.500	.809	b , b	.618	.809	a , c	.764	.382	b , a	.366
.500	.866	c , b	.577	.789	a , c	.732	.366	b , a	.352
.500	.750	a , b	.667	.889	b , c	.750	.500	a , b	.360
.500	.750	a , b	.667	.833	a , c	.800	.400	b , a	.362
.520	.853	c , b	.610	.880	c , c	.693	.360	a , a	.329
.532	.879	c , b	.605	.815	a , c	.742	.449	a , b	.338
.538	.836	b , b	.644	.899	c , c	.717	.386	a , a	.321
.544	.839	b , b	.648	.839	a , c	.772	.500	a , b	.357
.547	.876	c , b	.624	.859	b , c	.727	.397	a , a	.317
.555	.802	a , b	.692	.863	a , c	.802	.555	a , b	.368
.563	.852	b , b	.661	.885	b , c	.747	.421	a , a	.317
.570	.895	c , b	.637	.844	a , c	.755	.430	a , a	.307
.570	.815	a , b	.699	.926	c , c	.755	.430	a , a	.311
.586	.828	a , b	.707	.914	b , c	.773	.453	a , a	.313
.594	.871	b , b	.682	.871	a , c	.783	.466	a , a	.316
.618	.854	a , b	.724	.894	a , c	.809	.500	a , a	.314

Table 8.3 *The 47 section-invariant numbers with* $|\mathcal{U}| = 3$ *and* $b' > b$*;* L_v *(resp.* R_v*) stands for* $L(v, v')$ *(resp.* $R(v, v')$*)*

8.3, we obtain for the second line of the table $a = (b' - b)/(1 - b)$, $c = b/b'$, *and for the third line* $b = (c - c')/(1 - c')$, $c = c'/c$. *Solving for roots in* $[0, 1]$ *we get* $b = 1 - \psi \approx 0.4302$ *where* ψ *is defined as in Table 8.1(d), and* $a = \psi(1 - \psi) \approx 0.2451$, $c = 1 - \psi + \psi^2 \approx 0.7549$, $a' = b' = c' = \psi \approx 0.5698$. *Note that* b *is the solution of the equation* $b^2 = (1 - b)^3$, *a satisfies the equation* $a^3 - 4a^2 + 5a - 1 = 0$ *and the equation for* c *is* $c^3 + c^2 - 1 = 0$.

Families of algorithms are far more rich when $|\mathcal{U}| > 3$. Many

v	v'	$L(v,v')$	$R(v,v')$
a	a'	b	b
b	b'	a	c
c	c'	b	c

Table 8.4 *Example of section-invariant numbers with* $|\mathcal{U}| = 3$

structures have no solution, but some give a continuum of solutions, as illustrated by the following example.

Example 8.2 *The structure is given by Table 8.5. The values in this table are as follows:*

$$a = 2 - \tfrac{1}{c}, \quad a' = \tfrac{2c-1}{c^2}, \quad b = 1 - c, \quad b' = c,$$
$$d = \tfrac{1}{c} - 1, \quad d' = (\tfrac{1}{c} - 1)^2, \quad c' = 1 - c,$$

and c can be considered as a free parameter, with $1/2 < c < \varphi$ to guarantee $0 < a < b < c < d < 1$

v	v'	$L(v,v')$	$R(v,v')$
a	a'	a	c
b	b'	a	d
c	c'	a	d
d	d'	b	d

Table 8.5 *Example of section-invariant numbers with* $|\mathcal{U}| = 4$

A simple example with $m = 5$ is as follows.

Example 8.3 *The structure is given by Table 8.6, where* $a = 1/4$, $b = 1/3$, $c = 1/2$, $d = 2/3$, $f = 3/4$.

The following example shows how to construct arbitrarily large set $\mathcal{U}$.

Example 8.4 *Let* $s_n = \sum_{r=0}^{n} x^r$ *and consider the structure given in Table 8.7.*

The way to "close" the system is to set the term $1 - x$ equal to one term in column 1. For example:

v	v'	$L(v,v')$	$R(v,v')$
a	c	b	c
b	c	a	d
c	b	a	d
d	c	b	f
f	c	c	d

Table 8.6 *Example of section-invariant numbers with* $|\mathcal{U}| = 5$

v	v'	$L(v,v')$	$R(v,v')$
x	x^2	$x/(1+x)$	x
$\vdots$	$\vdots$	$\vdots$	$\vdots$
x^n	x^{n+1}	x^n/s_n	x
x^{n+1}	x^n	x^n/s_n	x
x^n/s_n	x^n	x^{n+1}	$1/s_n$
$1/s_n$	x^n/s_n	$1-x$	x^n
$\vdots$	$\vdots$	$\vdots$	$\vdots$

Table 8.7 *Example of section-invariant numbers with* $|\mathcal{U}|$ *arbitrarily large*

(i) $1 - x = x^n$, *which gives* $x = 1/2$ *when* $n = 1$, $x = \varphi$ *(Golden Section) when* $n = 2$, *etc.*

(ii) $1 - x = x^n/s_n$, *that is,* $x^{n+1} + x^n - 1 = 0$, *which gives* $x = \varphi$ *when* $n = 1$, *etc.*

The example in Table 8.8 was constructed by putting $1 - x = x^n$ *with* $n = 1$, *that is,* $x = 1/2$, *and allowing terms up to* $n + 1 = 3$.

Example 8.5 *In this example,* $m = 2(K+1)$ *and*

$$\begin{aligned}\mathcal{U} &= \{v_{(k,0)}, v_{(k,1)}, k = 0, \ldots, K\},\\ \mathcal{U}' &= \{v'_{(k,0)}, v'_{(k,1)}, k = 0, \ldots, K\},\end{aligned}$$

with, for $k \leq K$,

$$v_{(k,0)} = \frac{1}{1+2^k}, \quad v_{(k,1)} = \frac{2^k}{1+2^k},$$

v	v'	$L(v,v')$	$R(v,v')$
1/8	1/4	1/7	1/2
1/7	1/4	1/8	4/7
1/4	1/2	1/3	1/2
1/3	1/2	1/4	2/3
1/2	1/4	1/3	1/2
4/7	1/7	1/2	1/4
2/3	1/3	1/2	1/2

Table 8.8 *Example of section-invariant numbers with $|\mathcal{U}|$ arbitrarily large*

for $k < K$,

$$v'_{(k,0)} = \frac{v_{(k,0)}}{2}, \quad v'_{(k,1)} = \frac{1 + v_{(k,1)}}{2},$$

and

$$v'_{(K,0)} = \frac{1 + 2^{K-1}}{1 + 2^K}, \quad v'_{(K,1)} = \frac{2^{K-1}}{1 + 2^K}.$$

The structure is therefore given by Table 8.9, where $0 \le k \le K-1$.

v	v'	$L(v,v')$	$R(v,v')$
$v_{(k,0)} = \frac{1}{1+2^k}$	$v'_{(k,0)} = \frac{v_{(k,0)}}{2}$	$v_{(k+1,0)}$	$v_{(0,1)}$
$v_{(k,1)} = \frac{2^k}{1+2^k}$	$v'_{(k,1)} = \frac{1+v_{(k,1)}}{2}$	$v_{(0,0)}$	$v_{(k+1,1)}$
$\frac{1}{1+2^K}$	$\frac{1+2^{K-1}}{1+2^K}$	$v_{(0,0)}$	$v_{(K-1,0)}$
$\frac{2^K}{1+2^K}$	$\frac{2^{K-1}}{1+2^K}$	$v_{(K-1,1)}$	$v_{(0,1)}$

Table 8.9 *Example of section-invariant numbers with $|\mathcal{U}| = 2K + 2$*

References

J. Aaronson. *An Introduction to Infinite Ergodic Theory.* Am. Math. Soc., Providence, Rhode Island, 1997.

R.L. Adler. Geodesic flows, interval maps, and symbolic dynamics. In T. Bedford, M. Keane, and C. Series, editors, *Ergodic Theory, Symbolic Dynamics and Hyperbolic Spaces*, pages 93–123. Oxford Science Publications, Oxford, 1991.

H. Akaike. On a successive transformation of probability distribution and its application to the analysis of the optimum gradient method. *Ann. Inst. Statist. Math. Tokyo*, 11:1–16, 1959.

T. Bedford, M. Keane, and C. Series, editors. *Ergodic Theory, Symbolic Dynamics and Hyperbolic Spaces.* Oxford University Press, Oxford, 1991.

M. Berger. *Géométrie.* CEDIC/Nathan, Paris, 1979.

R.G. Bland, D. Goldfarb, and M.J. Todd. The ellipsoid method: a survey. *Operations Research*, 29(6):1039–1091, 1981.

R. Bowen. Invariant measures for Markov maps of the interval. *Commun. Math. Phys.*, 69:1–17, 1979.

I. Cornfeld, S.V. Fomin, and Ya. G. Sinai. *Ergodic Theory.* Springer, New York, 1982.

K. Dajani and C. Kraaikamp. Generalization of a theorem of Kusmin. *Mh. Math.*, 118:55–73, 1994.

L. Danzer, B. Grünbaum, and V. Klee. Helly's theorem and its relatives. In V.L. Klee, editor, *Proceedings of Symposia in Pure Mathematics, Vol. VII: Convexity*, pages 101–180. Am. Math. Soc., Providence, Rhode Island, 1963.

V. Drobot. Uniform partitions of an interval. *Trans. Amer. Math. Soc.*, 268(1):151–160, 1981.

D.Z. Du and F.K. Hwang. *Combinatorial Group Testing and its Applications.* World Scientific, Singapore, 1993.

J.G. Ecker and M. Kupferschmid. A computational comparison of the ellipsoidal algorithm with several nonlinear programming algorithms. *SIAM J. Control and Optimization*, 23(5):657–674, 1985.

Y.P. Evtushenko. Numerical methods for finding the global extrema of a nonuniform mesh. *USSR Computing Machines and Mathematical Physics*, 11:1390–1403, 1971.

V.V. Fedorov. *Theory of Optimal Experiments.* Academic Press, New York, 1972.

V.V. Fedorov and P. Hackl. *Model-Oriented Design of Experiments.* Springer, Berlin, 1997.

P. Flajolet and B. Vallée. Continued fraction algorithms, functional operators, and structure constants. *Theoretical Computer Science*, 194(1–2):1–34, 1998.

G.E. Forsythe. On the asymptotic directions of the s-dimensional optimum gradient method. *Numerische Mathematik*, 11:57–76, 1968.

F.R. Gantmacher. *Théorie des Matrices.* Dunod, Paris, 1966.

J. Hale and H. Koçak. *Dynamics and Bifurcations.* Springer-Verlag, Heidelberg, 1991.

J.L. Jensen. Chaotic dynamical systems with a view towards statistics: a review. In O.E. Barndorff-Nielsen, J.L. Jensen, and W.S. Kendall, editors, *Networks and Chaos— Statistical and Probabilistic Aspects*, pages 201–250. Chapman & Hall, London, 1993.

F. John. Extremum problems with inequalities as subsidiary conditions. In J. Moser, editor, *F. John Collected Papers, Vol. 2*, pages 543–560. Birkhäuser, Boston, 1985. Originally published in *Courant Anniversary Volume, 1948, pp. 187–204.*

P. Kargaev and A.A. Zhigljavsky. Asymptotic distribution of the distance function to the Farey points. *J. of Number Theory*, 65(1):130–149, 1997.

M.S. Keane. Ergodic theory and subshifts of finite type. In T. Bedford, M. Keane, and C. Series, editors, *Ergodic Theory, Symbolic Dynamics and Hyperbolic Spaces*, pages 35–70. Oxford Science Publications, Oxford, 1991.

J.E. Kelley. The cutting plane method for solving convex programs. *SIAM Journal*, 8:703–712, 1960.

L.G. Khachiyan. A polynomial algorithm in linear programming. *Doklady AkademiïaNauk SSSR*, 244:1093–1096, 1979. Translated into English in *Soviet Mathematics Doklady*, 20, 191–194.

L.G. Khachiyan and M.J. Todd. On the complexity of approximating the maximal inscribed ellipsoid for a polytope. *Math. Programming*, A61(2):137–159, 1993.

J. Kiefer. Sequential minimax search for a maximum. *Proc. Am. Math. Soc.*, 4:502–506, 1953.

J. Kiefer. Optimum sequential search and approximation methods under minimum regularity assumptions. *J. Soc. Indust. Appl. Math.*, 5(3):105–136, 1957.

J. Kiefer and J. Wolfowitz. Optimum designs in regression problems. *Annals of Math. Stat.*, 30:271–294, 1959.

J. Kiefer and J. Wolfowitz. The equivalence of two extremum problems. *Canadian Journal of Mathematics*, 12:363–366, 1960.

D.E. Knuth. The distribution of continued fraction approximations. *Journal of Number Theory*, 19:443–448, 1984.

M. Krasnoselskii. *Positive Solutions of Operator Equations.* P. Noordhoff, Groningen, 1964.

J.C. Lagarias. Number theory and dynamical systems. *Proccedings of Symposia in Applied Mathmetics*, 46:35–72, 1992.

A.Ju. Levin. On an algorithm for the minimization of convex functions. *Soviet Math. Dokl.*, 6:286–290, 1965.

D. Lind and B. Marcus. *An Introduction to Symbolic Dynamics and Coding.* Cambridge University Press, Cambridge, 1995.

D.G. Luenberger. *Introduction to Linear and Nonlinear Programming.* Addison-Wesley, Reading, Massachusetts, 1973.

R. Mañé. *Ergodic Theory and Differentiable Dynamics.* Springer, Berlin, 1987.

D.H. Mayer. Continued fractions and related transformations. In T. Bedford, M. Keane, and C. Series, editors, *Ergodic Theory, Symbolic Dynamics and Hyperbolic Spaces*, pages 175–222. Oxford Science Publications, Oxford, 1991.

M. Minoux. *Programmation Mathématique, Théorie et Algorithmes, vol. 1 & 2.* Dunod, Paris, 1983.

J. Nocedal, A. Sartenaer, and C. Zhu. On the accuracy of nonlinear optimization algorithms. Technical Report Nov. 98, ECE Department, Northwestern Univ., Evanston, Il 60208, 1998.

J.H. O'Geran, H.P. Wynn, and A.A. Zhiglyavsky. Search. *Acta Applicandae Mathematicae*, 25:241–276, 1991.

J.H. O'Geran, H.P. Wynn, and A.A. Zhiglyavsky. Mastermind as a test-bed for search algorithms. *Chance*, 6:31–37, 1993.

E. Ott. *Chaos in Dynamical Systems.* Cambridge University Press, New York, 1993.

A. Pázman. *Foundations of Optimum Experimental Design.* VEDA (co-pub. Reidel, Dordrecht), Bratislava, 1986.

K. Petersen. *Ergodic Theory.* Cambridge University Press, Cambridge, 1983.

J.D. Pinter. *Continuous and Lipschitz Optimization: Algorithms, Implementations and Applications.* Kluwer, Dordrecht, 1995.

L. Pronzato and E. Walter. Volume-optimal inner and outer ellipsoids, with application to parameter bounding. In *Proc. 2nd European Control Conf.*, pages 258–263, Groningen, June 1993.

L. Pronzato and E. Walter. Minimal-volume ellipsoids. *Int. Journal of Adaptive Control and Signal Processing*, 8:15–30, 1994.

L. Pronzato and E. Walter. Minimum-volume ellipsoids containing compact sets: application to parameter bounding. *Automatica*, 30(11):1731–1739, 1994.

L. Pronzato and E. Walter. Volume-optimal inner and outer ellipsoids.

In M. Milanese, J.-P. Norton, H. Piet-Lahanier, and E. Walter, editors, *Bounding Approaches to System Identification*, pages 119–138. Plenum, London, 1996.

L. Pronzato, H.P. Wynn, and A.A. Zhigljavsky. Dynamic systems in search and optimisation. *CWI Quarterly*, 8(3):201–236, 1995.

L. Pronzato, H.P. Wynn, and A.A. Zhigljavsky. Stochastic analysis of convergence via dynamic representation for a class of line-search algorithms. *Combinatorics, Probability & Computing*, 6:205–229, 1997.

L. Pronzato and A.A. Zhigljavsky. On average-optimal quasi-symmetrical univariate optimization algorithms. In W.G. Müller, H.P. Wynn, and A.A. Zhigljavsky, editors, *Model-Oriented Data Analysis III, Proceedings MODA3, St Petersburg, May 1992*, pages 269–278. Physica Verlag, Heidelberg, 1993.

A. Rényi. On measures of entropy and information. In *Proc. 4th Berkeley Symp. on Math. Statist. and Prob.*, pages 547–561, 1961.

A. Rényi. On the foundations of information theory. *Rev. Inst. Internat. Stat.*, 33:1–14, 1965.

A.H.G. Rinnooy-Kan and G.T. Timmer. Global optimization. In G.L. Nemhauser *et al.*, editor, *Handbooks in OR & MS, vol. 1*, pages 631–662. Elsevier, Amsterdam, 1989.

A.M. Rockett and P. Szüsz. *Continued Fractions*. World Scientific, Singapore, 1992.

D. Ruelle. *Chaotic Evolution and Strange Attractors*. Cambridge University Press, New York, 1989.

J. Schoisengeier. An asymptotic expansion for $\sum_{n<N}\{n\alpha+\beta\}$. In H. Hlawka and R.F. Tichy, editors, *Number-Theoretic Analysis*, pages 199–205. Springer-Verlag, Heidelberg, 1990.

M. Schroeder. *Fractals, Chaos, Power Laws: Minutes from an Infinite Paradise*. W.H. Freeman and Company, New York, 1991.

N.Z. Shor. *Minimization Methods for Non-Differentiable Functions*. Springer, Berlin, 1985.

N.Z. Shor and O.A. Berezovski. New algorithms for constructing optimal circumscribed and inscribed ellipsoids. *Optimization Methods and Software*, 1:283–299, 1992.

R. Sibson. Discussion on a paper by H.P. Wynn. *Journal of Royal Statistical Society*, B34:181–183, 1972.

S.D. Silvey. *Optimal Design*. Chapman & Hall, London, 1980.

S.P. Tarasov, L.G. Khachiyan, and I.I. Erlich. The method of inscribed ellipsoids. *Soviet Math. Dokl.*, 37(1):226–230, 1988.

D.M. Titterington. Algorithms for computing D-optimal designs on a finite design space. In *Proc. of the 1976 Conference on Information Science and Systems*, pages 213–216, Baltimore, 1976. Dept. of Electronic Engineering, John Hopkins University.

D.M. Titterington. Estimation of correlation coefficients by ellipsoidal trimming. *Journal of Royal Statistical Society*, C27(3):227–234, 1978.

B. Vallée. Opérateurs de Ruelle-Mayer généralisés et analyse en moyenne des algorithmes d'Euclide et de Gauss. *Acta Arithmetica*, 81(2):101–144, 1997.

E. Walter and L. Pronzato. Characterizing sets defined by inequalities. In M. Blanke and T. Söderström, editors, *Prep. 10th IFAC/IFORS Symposium on Identification and System Parameter Estimation*, volume 2, pages 15–26, Copenhagen, July 1994. Danish Automation Society.

E. Walter and L. Pronzato. *Identification of Parametric Models from Experimental Data.* Springer, Heidelberg, 1997.

M.H. Wright. The interior-point revolution in constrained optimization. Technical Report 98–4–09, Computing Sciences Research Center, Bell Laboratories, Murray Hill, New Jersey 07974, 1998.

H.P. Wynn. The sequential generation of D-optimum experimental designs. *Annals of Math. Stat.*, 41:1655–1664, 1970.

H.P. Wynn and A.A. Zhigljavsky. Chaotic behaviour of search algorithms: introduction. In W.G. Müller, H. P. Wynn, and A.A. Zhigljavsky, editors, *Model-Oriented Data Analysis III, Proceedings MODA3, St Petersburg, May 1992*, pages 199–211. Physica Verlag, Heidelberg, 1993.

H.P. Wynn and A.A. Zhigljavsky. The theory of search from a statistical viewpoint. *Test*, 3:1–45, 1994.

H.P. Wynn and A.A. Zhigljavsky. Achieving the ergodically optimal convergence rate for a one-dimensional minimization problem. *Journal of Complexity*, 11:196–2, 1995.

A.A. Zhigljavsky. *Theory of Global Random Search.* Kluwer, Dordrecht, 1991.

A.A. Zhigljavsky and M.V. Chekmasov. Comparison of independent, stratified and random covering sample schemes in optimization problems. *Math. Comput. Modelling*, 23(8–9):97–110, 1996.

Author index

Subject index